Discrete Mathematics
for
Every Highschooler

Alexander V. Sadovsky

ii

Illustrations by the author.

Typeset and formatted using LaTeX. Figures hand-drawn and created using Microsoft PowerPoint and processed using GIMP.

Contents

Foreword

George Orwell opens his article *Politics and the English language* (1946) by saying,

> Most people who bother with the matter at all would admit that the English language is in a bad way, but it is generally assumed that we cannot by conscious action do anything about it.

The same is true of the K-12 mathematical education in the U.S.A.: it is in such a "bad way" that the external symptom is the mass "math-phobia." My evidence for this statement comes from three sources, my own academic experience, the American popular culture, and the market of textbooks in elementary mathematics. The first two of these show that, to students here (children and adults alike), mathematics generally means memorization of long and disconnected lists of useless formulas, leading to frustration and pain. The publishing market responds to this with popular books offering therapy for the traumatic shock of "math" and advice how to "survive it." This, unfortunately, only further spreads the misconception that mathematics is there to force the mind into unnatural and painful contortions.

Two likely immediate causes of this mass phobia are the following two doctrines, which I have come to regard as completely false:

- Mathematics is about numbers and arithmetic (rather than ideas and applications).

- Problem solving is about "breaking it down into steps."

Besides the mentioned "math-phobia," the above two doctrines have numerous other consequences. The one addressed in this book is the situation of many students whose major will involve one or more of the following: set theory, combinatorics, probability, statistics. These students' situation is frequently as follows. The time allocated for discrete mathematics by most college curricula in the U.S. is far from sufficient. (Is it because the material is considered "too hard" for highschool and held off until college? At what minimal age can we expect someone to understand lists? Cross-referencing two lists? Concatenating two lists with no duplicates? Assigning informative labels to items on a list? Describing all possible customer behaviors in a given setting? These practical ideas are what stands behind the mathematical concepts of *intersection of sets*, *union of sets*, *function*, and *sample space*.) In the computer science major alone, most curricula push the student from basic set theory to advanced discrete algorithms inside of two or three academic terms. The problem, however, is by no means unique to computer science. Business and social studies are accompanied by a similar race through probability and statistics. Such rapid initiation into the subject leaves no time for the student to assimilate the ideas and to become proficient in their use. Allocating more time, on the other hand, is often economically infeasible, especially for universities subsidized by the state. As a result, the student has been made to skip a firm footing in the basics: earlier (high school), these were never taught in detail because it was "too early," and now (college) they are not taught because there is not enough time in the curriculum.

The solution, in my opinion, is an earlier introduction of the basics. A first elementary exposition to a mathematical subject is known to have done wonders in the student's preparation for the deeper study. But the gain is not just in that particular subject. All mathematics is connected. Knowledge of discrete mathematics helps with other subjects: exposure to the concepts of sets and functions in broader context enables the student to grasp much more easily the word problems (read: applications) of algebra, trigonometry, and calculus. Such an exposition to the basics of discrete mathematics is the content of this book.

Basic discrete mathematics is age-appropriate as early as mid- to upper teens, if not earlier, and makes, for reasons stated above, a critical difference in the success of every college career in every field of engineering and natural science.

Mathematics is experimental[1]: working (experimenting) with concrete examples is key. To aid the experimentation, some of the examples and exercises in this book rely on the use of a spreadsheet program. Unless noted otherwise, the spreadsheet syntax used is suitable for most standard spreadsheets, including those provided by Google Documents, Microsoft Excel, Mac OS, and Linux's OpenOffice and Staroffice.

The author's warmest thanks go to the following individuals, whose advice and technical expertise proved a valuable asset: A. Berger, M. Dynin, K. Sadovsky, D. Faradjev, D. Khidekel, Y. Yuryev.

---

[1]For a well-argued presentation of this view, see the article cited under reference [1]. Its author, the late Vladimir Arnol'd, was a world-class mathematician and educator, whose teaching encompassed students as young as 4.

How to use this book

Hints, answers, and solutions

Harald Bohr[2] is believed to have said, "While mathematics does not teach us to think correctly, it teaches us how easy it is to think incorrectly." Anyone solving a problem in science or engineering, whether a novice or a seasoned researcher, needs also ways to validate the obtained solution. Finding such ways is a challenge in itself and should not be put to the beginner, who has enough to handle as it is. To give the reader an independent source of validation, I have provided hints, solutions, and answers to selected exercises in appendix D.

Numeration of formulas, examples, and exercises

Some of the formulas are numbered and, elsewhere in the book, referred to by the number. The numeration is continuous within each chapter; for example, the 3rd numbered formula in chapter 2 is accompanied by the label (2.3), regardless of the section in which it occurs. The same numeration scheme, although typeset differently, applies to the examples and exercises.

---

[2]Harald Bohr (1887 - 1951) was a Danish mathematician and football player.

References to other literature

The section called *Bibliography*, immediately before the appendices, contains numbered sources (mainly, books and articles), to which references are made throughout this book. Thus, for example, [12] refers to Frederic Mosteller's book entitled *Fifty Challenging Problems in Probability with Solutions*. Some of the books are listed without a specific edition: this means, any edition of the book will do for the purposes for which I cite it.

The Index

If you encounter a term whose definition, given earlier, you forgot and have trouble finding, it can be looked up using the Index, located at the very end of the book. Looking things up is an essential study skill, whether you intend to go into science, accounting, medicine, law, plumbing, or business.

What not to do, and what to do instead

Mechanical memorization should not be used long term. Instead of being satisfied with just remembering a formula (a common path of least resistance), understand how to derive it from first principles[3]. While a greater investment at first, this approach makes you independent of how good your memory is and removes the otherwise huge risk of misusing your mathematics. In other words, the payoff from the investment is a lasting education rather than dust and ashes of facts once memorized.

---

[3]A brilliant illustration of how there is no need to memorize the formulas for expanding $(a + b)^2$ and $(a + b)^3$ is given in the book [3].

Chapter 1

Sets and functions

I read the prescription. It ran:

1 lb. beefsteak, with 1 pt. bitter beer every 6 hours.

1 ten-mile walk every morning.

1 bed at 11 sharp every night.

And don't stuff up your head with things you don't understand.

Jerome K. Jerome

1.1 Sets and their elements

The term *set* will be used intuitively, i.e. without a formal definition, to mean a collection of objects. The objects that constitute the set are said to be contained in the set, and are called, *the elements* of the set. If a set X consists of the objects $a, b, c \ldots$, then one uses the notation $X = \{a, b, c, \ldots\}$. A set cannot be its own element.

Some examples of sets are:

- $\{\clubsuit, \spadesuit, \heartsuit, \diamondsuit\}$

- the set of the letters that constitute the English alphabet

- the set $\{-1, 5, 3, 0 - 2.3\}$

7

- the set of all negative integers; i.e., the set

$$\{n : n \text{ is a negative integer}\}$$

- the set of all points in a plane

- the set of points in a plane equidistant from a given point (i.e., a circle[1])

- the set of points in a plane within a given distance from a given point (i.e., a circular disc)

- the set of all cities connected to a given one by a road

- the set of all employee records in a companys database

- a shopping list

Since listing an item on a shopping list twice is redundant, the last example illustrates that a set does not contain "multiple copies" of the same element; e.g., the sets $\{1, 2, 2\}$ and $\{1, 2\}$ are one and the same set. The mathematics corresponding to this intuition will be given below.

A set has no "intrinsic order" for its elements.

Example 1.1 *The sets $\{1, 2, 7\}$ and $\{2, 7, 1\}$ are one and the same set.*

To distinguish order, another construction (*ordered pairs*, defined below) is used.

1.2   The empty set

The following axiom will be accepted as true throughout this book.

Axiom. *There exists one, and only one, set with no elements.*

The set mentioned in the above axiom is called *the empty set*, and is denoted $\emptyset$. All other sets are said to be nonempty.

---

[1]The definition and construction of a circle are given in reference [7].

1.3 Subsets

A set Y is said to be a *subset* of a set X if every element of Y is also an element of X. Two sets are said to be *equal* if they are subsets of each other, and to be *distinct (or unequal)* if not equal. Equality and inequality of sets are denoted, respectively, by the symbols $=$ and $\neq$. This means that two sets are equal if and only if they have exactly the same elements.

Example 1.2 *A statement of equality for two specific sets is made in example 1.1.*

Example 1.3 *The set $\{\clubsuit, \heartsuit\}$ is a subset of the set $\{\heartsuit, \spadesuit, \clubsuit\}$.*

Example 1.4 $\{\clubsuit, \heartsuit, \spadesuit\} = \{\heartsuit, \spadesuit, \clubsuit\}$ *This equality is verified by showing that each set is a subset of the other.*

Example 1.5 $\{\clubsuit, \heartsuit, \spadesuit\} \neq \{\diamondsuit, \clubsuit, \heartsuit\}$

Example 1.6 *Examining the sets $\{1, 2, 2\}$ and $\{1, 2\}$, one finds that each is a subset of the other, hence $\{1, 2, 2\} = \{1, 2\}$. A similar examination establishes that $\{1, 2, 7\} = \{2, 7, 1\}$ (i.e., a set does not ascribe any intrinsic order to its elements).*

Example 1.7 *The statement "Every horse is a mammal" can be put, "The set of all horses is a subset of the set of all mammals."*

Example 1.8 *The empty set $\emptyset$ is a subset of every set.*

Example 1.9 *Every set (including $\emptyset$) is its own subset.*

Exercise 1.1 *Suppose A is a subset of B, and B is a subset of C. Is A necessarily a subset of C?*

1.4 Set operations

Given sets X and Y,

- the set consisting of all elements common to X and Y is called, *the intersection of X and Y and is denoted $X \cap Y$*:

$$X \cap Y$$

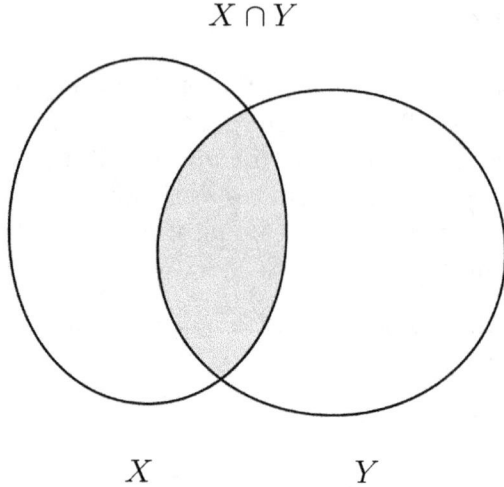

X Y

- the set consisting of all elements that lie either in X or in Y or in both is called, *the union of X and Y* and is denoted $X \cup Y$:

$$X \cup Y$$

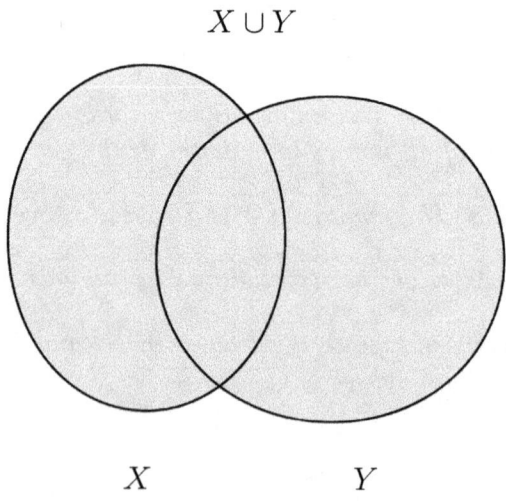

X Y

- the set consisting of all elements in X that are not in Y is called, *the complement of Y in X*, and is denoted $X \setminus Y$:

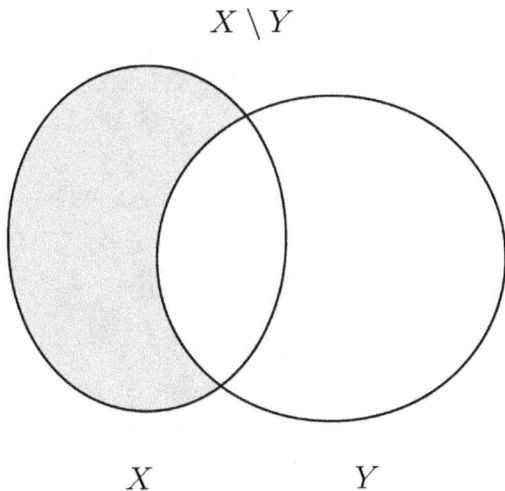

$$X \setminus Y$$

$$X \qquad\qquad Y$$

The drawings used in the above definition are called *Venn's diagrams*[2].

Exercise 1.2 *Two sets are said to be* disjoint *if they have no common elements (i.e., if their intersection is the empty set). For each of the following pairs of sets, determine whether the sets are disjoint.*

(a) $\{\clubsuit, \spadesuit, \diamondsuit\}$ *and* $\{\heartsuit\}$

(b) $\{1, 2, 3\}$ *and* $\{4, 5, 6\}$

(c) $\{3, 1, 2\}$ *and* $\{4, 7, 5\}$

(d) $\{3, 1, 2\}$ *and* $\{4, 2, 7\}$

(e) $\{1, 2, 3, \heartsuit\}$ *and* $\{4, 5, \clubsuit, \spadesuit\}$

(f) $\{0, 1, 2, 3, \dots, 93\}$ *and* $\{16, 18, 20, 22, \dots, 96\}$

Exercise 1.3 *If X is any subset of a set Q, denote the set $Q \setminus X$ (the complement of X in Q) by X^c (the c stands for complement). The operation of complement takes precedence over the other set operations, e.g. $X \cup Y^c$ means $X \cup (Y^c)$. (Similarly for $\cap$.)*

---

[2]John Venn FRS (1834 - 1923) was a British logician and philosopher.

Suppose R and S are subsets of Q. Use Venn's diagrams to convince yourself that the following formulas hold:

$$(R \cap S)^c = R^c \cup S^c, \quad (R \cup S)^c = R^c \cap S^c \tag{1.1}$$

Formulas (1.1) are called, *DeMorgan's Laws*[3].

1.5 Membership functions

Suppose Q is a set, and R a subset of Q. To each element x of Q, associate the value 1 if x is an element of R, and the value 0 otherwise. The rule that associates these values is called, *the membership function of R*, and is denoted m_R.

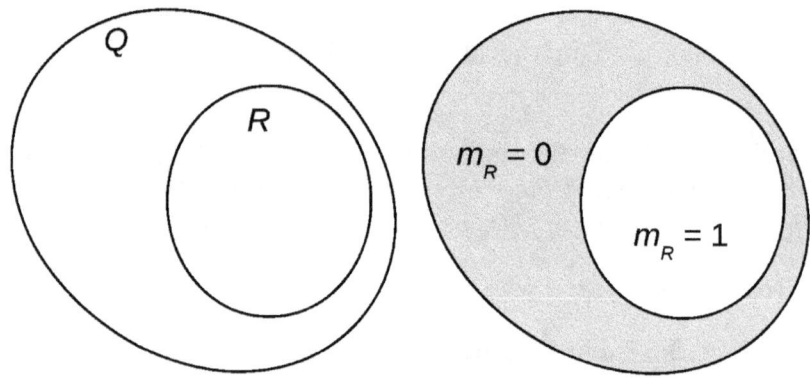

The value associated to x is written $m_R(x)$. Thus,

$$m_R(x) = \begin{cases} 1 & \text{if } x \text{ is an element of } R \\ 0 & \text{otherwise} \end{cases}$$

Example 1.10 *If $Q = \{a, b, c, d, e, f\}$ and $R = \{b, d, e, f\}$, then m_R is described by the table*

x	a	b	c	d	e	f
$m_R(x)$	0	1	0	1	1	1

---

[3]Augustus DeMorgan (1806 - 1871) was a British mathematician and logician.

In a spreadsheet with the columns marked A, B, C, D, E, F, *etc., one can record* m_R *in the first row of the sheet by mimicking the above table, as follows:*

	A	B	C	D	E	F	
row 1	0	1	0	1	1	1	...
2							...
.							...
.							...
.							...

Another example is the empty set, $\emptyset$, *regarded as a subset of* Q. *The membership function of* $\emptyset$ *is always identically zero; recording it in row 2 in the spreadsheet shown above would give*

	A	B	C	D	E	F	
row 1	0	1	0	1	1	1	...
2	0	0	0	0	0	0	...
.							...
.							...
.							...

Exercise 1.4 *Membership functions provide a way to carry out set operations using a spreadsheet (and, generally, a computer). If R and S are subsets of Q, then the membership function of $R \cap S$ is the product of the membership functions of R and S:*

$$m_{R \cap S}(x) = m_R(x)m_S(x)$$

This equality follows from the fact that if two numbers each have value 0 or 1, then their product equals 1 if both numbers are 1, and equals zero otherwise.

(a) *Use a spreadsheet to test the latter equality on the sets $Q = \{a, b, c, d, e, f\}$, $R = \{b, d, e, f\}$, $S = \{b, c, d, e\}$.*

(b) *Express the membership function of $R \cup S$ in terms of the membership functions of R and S.*

(c) *Express the membership function of* $Q \setminus R$ *in terms of the membership function of* R.

Exercise 1.5 *If* R *and* S *are subsets of* Q, *and* x *an element of* Q, *one can ask whether the statement*

$$\text{Object } x \text{ is an element of } R. \tag{1.2}$$

is true or false. Denote statement (1.2) by $r(x)$. *It will be true or false for a given* x, *accordingly as the value* $m_R(x)$ *is one or zero. In the former case, we will write*

$$r(x) = T;$$

in the latter,

$$r(x) = F$$

The symbols T *and* F *are called, respectively,* truth *and* falsity, *and are collectively known as* truth values. *A declarative statement whose truth value can in principle be decided*[4] *is called, a* proposition. *A proposition with truth value* T *is said to be* true; *one with truth value* F, *to be* false. *The study of truth values of propositions is called,* propositional logic.

Analogously to proposition $r(x)$, *defined by (1.2), define proposition* $s(x)$ *by*

$$\text{Object } x \text{ is an element of } S. \tag{1.3}$$

(a) *The proposition*

$$\text{Propositions } r(x) \text{ and } s(x) \text{ are both true.} \tag{1.4}$$

will be true if and only if $r(x)$ *and* $s(x)$ *are both true (for the same* x). *Proposition (1.4) is called,* the conjunction *of propositions* $r(x)$ *and* $s(x)$, *and is denoted*

$$(r \wedge s)(x)$$

---

[4]Believe it or not, there exist statements that can be mathematically proved to be undecidable. Such proofs are by no means elementary; e.g., see reference [5].

If x is an element of Q, fill the appropriate truth values in the rightmost column of the table

$r(x)$	$s(x)$	$(r \wedge s)(x)$
F	F	
F	T	
T	F	
T	T	

(the latter table is called, the truth table *of proposition* $(r \wedge s)$*) and state an analogy between conjunction* $\wedge$ *and intersection* $\cap$.

(b) *The proposition*

At least one of the propositions $r(x)$ and $s(x)$ is true.
$$(1.5)$$
is called, the disjunction *of the propositions* $r(x)$ *and* $s(x)$*, and is denoted*

$$(r \vee s)(x)$$

Fill the truth table

$r(x)$	$s(x)$	$(r \vee s)(x)$
F	F	
F	T	
T	F	
T	T	

Which set operation is disjunction analogous to, and how?

(c) *The proposition*

Proposition $r(x)$ is false. (1.6)

is called, the negation *of proposition* $r(x)$*, and is denoted*

$$\neg r(x)$$

Fill the truth table

$$\begin{array}{c|c} r(x) & \neg r(x) \\ \hline F & \\ T & \end{array}$$

Which set operation is negation analogous to, and how?

(**d**) *Conjunction* $\wedge$, *disjunction* $\vee$, *and negation* $\neg$ *are called,* logical operations, *as they are used to make new propositions from existing ones. State the version of De-Morgan's laws that describes propositions and logical operations (analogously to laws for subsets and set operations).*

1.6   Ordered pairs and Cartesian products

The idea of a coordinate plane is said to have come to Descartes[5] as he was watching a fly crawl across the ceiling. He realized that the fly's position could be specified completely by a pair of distances, measured from the corner of the ceiling (the origin) along its two perpendicular edges (to distinguish them, call them *edge 1* and *edge 2*). Whether true or not, this anecdote illustrates the idea of the Cartesian coordinate system, named after Descartes, and its essence: the ordered pair. "Ordered," because 3 feet along edge 1 and 4 feet along edge 2 is not the same as vice versa: the two flies pictured below occupy two different positions (one division along an edge is one foot long).

---

[5]René Descartes (1596 - 1650) was a French mathematician, physicist, and philosopher.

edge 1

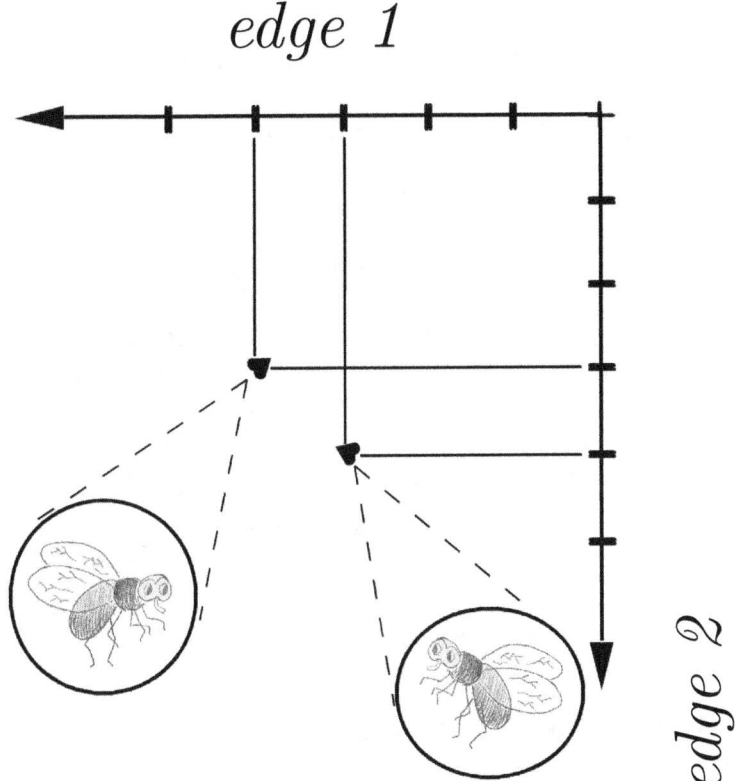

edge 2

In the subsequent centuries, and in many adjacent fields, there frequently came a need to pair objects other than numbers: to keep a record, in a class, of each student's letter grade (pairing the name with the grade); to label each registered car with a license plate; to investigate social- and business connections in a criminal case; to engineer a telephone grid for a city; to plan the routes of an airplane or a boat; to assess the cost of each business strategy considered.

Given objects x and y, and picking one of the two (say, x), one lists them thus

$$(x, y),$$

with the understanding that, if x and y are two distinct objects, then (x, y) and (y, x) are two distinct lists. Such a list is called, *an ordered pair*[6].

---

[6] Although this definition will suffice for our purposes, the reader may feel

Exercise 1.6 *In the Cartesian coordinate plane, graph the points* $(1, 2)$ *and* $(2, 1)$, *thus illustrating that they are distinct. (The reader who has not seen a coordinate line or a coordinate plane can get acquainted with them using reference [4]).*

Two ordered pairs (x_1, y_1) and (x_2, y_2) coincide if

$$x_1 = x_2 \text{ and } y_1 = y_2$$

Given sets X and Y, the set of *all* ordered pairs (x, y) such that x is an element of X and y an element of Y is called *the Cartesian product of X and Y*, and is denoted $X \times Y$.

Example 1.11 *If* X *is the set of all real numbers, then* $X \times X$ *is the Cartesian coordinate plane (i.e., the set of all ordered pairs of real numbers).*

Exercise 1.7 *For each pair of sets, calculate their Cartesian product:*

(a) $X = \{\heartsuit, \spadesuit, \diamondsuit\}, \quad Y = \{a, c, j\}$

(b) $X = \{\heartsuit, 1, a\}, \quad Y = \{a, a, c\}$

(c) $X = Y =$ *the set of all integers.*

(d) $X = \{a, b, c, d, ..., z\}, \quad Y = \emptyset$

(e) $X = Y = \emptyset$

---

dissatisfied that it is not rigorous. Indeed, how does such a list "know" which of the two objects "comes first?" Is there a definition that captures this intrinsically? There is. One rigorous definition of the *ordered pair* (x, y) is the set $\{\{x, 1\}, \{y, 2\}\}$ (a set of sets). This construction is "able to distinguish between x and y" in the following sense: if the objects x and y are distinct ($x \neq y$), then

$$\begin{aligned} (x, y) &= \{\{x, 1\}, \{y, 2\}\} \\ &\neq \{\{y, 1\}, \{x, 2\}\} \\ &= (y, x) \end{aligned}$$

i.e. $(x, y) \neq (y, x)$.

Analogously to the ordered pair, there are the *ordered triple*, the *ordered quadruple*, and so on, the emerging generic concept being that of *an ordered n-tuple*. With these, one defines, for instance, *the Cartesian product $A \times B \times C$ of the given sets A, B, C as the set consisting of all ordered triples

$$(a, b, c),$$

where a, b, c are, respectively, elements of A, B, C.

Exercise 1.8 *Consider the sets*

$$B = \{\texttt{wheat, rye, multigrain}\},$$

$$S = \{\texttt{ketchup, mayonnaise}\},$$

$$V = \{\texttt{tomato, onion, cucumber, avocado, lettuce}\},$$

$$C = \{\texttt{swiss, smoked, cheddar, feta}\}.$$

(a) *The Cartesian product $B \times S \times V \times C$ will list certain kinds of cheese sandwich that can be made with the given ingredients. Find five elements (ordered 4-tuples) of this Cartesian product.*

(b) *Each kind of cheese sandwich found in the previous part of the exercise allows only one kind of cheese, one vegetable, and one kind of spread. What sets and what Cartesian products can one form to allow more than one kind of cheese, vegetables, and spread on one sandwich? Hint: Look up the concept of* power set *in exercise 2.6, below.*

1.7 Functions

A *function from a set X to a set Y* is a rule that assigns to each element x of X exactly one element y of Y:

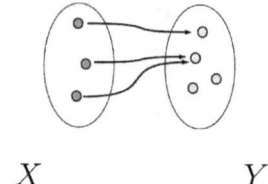

$$X \qquad\qquad Y$$

If y is assigned to x, one says that the function *maps* or *carries* x to y. (In the above figure, the arrows show which elements in X are mapped to which in Y. Notice that Y can have elements to which no element of X is mapped.)

Example 1.12   (a) *If X is a set of books, by assigning to each book its weight, one defines a function from X to the set of the real numbers. By assigning to each book the number of pages in it, one defines a function from X to the set of integers.*

(b) *If X is a set of a company's product models, the manufacturing engineer often defines the function that maps each model to its the minimal expected durability.*

(c) *In ecommerce, one often works with the function that maps each web link to the number of hits it has received over some period of time.*

(d) *If X is a list of shopping items, by specifying the needed quantity of each item one defines a function from X to a set of numbers.*

(e) *If X is the set of integers, the function that maps each integer x to 0 or to 1 accordingly as x is even or odd is called* parity.

(f) *Let $X = \{a, b, c\}$ and $Y = \{1, 2, 3\}$. Attempting to map a to 1, b to 2, c to 3, and b to 1, one fails to obtain a function from X to Y. By definition (the first sentence of this section), a function from X to Y must map each element of X to* **exactly one** *element of Y. This condition is violated by trying to map b both to 2 and to 1.*

(g) *Let $X = \{a, b, c\}$ and $Y = \{1, 2, 3\}$. Attempting to map a to 1, b to 2, with no other mappings, one fails to obtain a function from X to Y. By definition (the first sentence of this section), such a function must map* **each** *element of X to an element of Y. This condition is violated by failing to map c to any element of Y.*

(h) *Let X be the set of all real numbers, and $Y = X$. The function mapping each x in X to the square x^2 can be graphed in a coordinate plane. The graph is a curve known as a* quadratic parabola.

Exercise 1.9 *As is mentioned above, if X is a list of shopping items, then by specifying the needed quantity of each item one defines a function from X to a set Y that consists of some numbers. Does Y necessarily consist of integers only?*

If f is a function from X to Y and maps an element x of X to an element y of Y, then one writes

$$y = f(x)$$

Exercise 1.10 *Let $X = \{1, 2, 3, 4\}$, and let f be a function that maps every number in X to twice that number.*

(a) *Compute $f(1)$, $f(3)$ and $f(4)$.*

(b) *Is f a function from X to $Y = \{2, 4, 6, 8\}$?*

(c) *Is f a function from X to $Y = \{-1, 0, 1, 2, 3, 4, 5, 6, 7, 8, 9, 11\}$?*

(d) *Is f a function from X to $Y = \{2, 4, 5, 8\}$?*

1.8 1-to-1 Correspondences

A *1-to-1 correspondence* from a set X to a set Y is a function f from X to Y such that

- no two distinct elements of X are mapped to the same element of Y, and

- each element of Y has one of X mapped to it.

Exercise 1.11 *For each the following, determine whether it is a 1-to-1 correspondence f from $X = \{a, b, c\}$ to $Y = \{1, 2, 3\}$:*

(a) $f(a) = b$, $f(b) = 1$, $f(c) = 3$

(b) $f(a) = 3$, $f(b) = 2$, $f(c) = 2$

(c) $f(a) = 3$, $f(b) = 2$, $f(a) = 1$

(d) $f(a) = 1$, $f(b) = 2$

(e) $f(2) = a$, $f(3) = b$, $f(1) = c$

(f) $f(a) = 2$, $f(b) = 3$, $f(c) = 1$

Exercise 1.12   (a) *Does there exist a 1-to-1 correspondence from $\{a, b, c, d\}$ to $\{1, 2, 3\}$?*

(b) *Does there exist a 1-to-1 correspondence from $\{a, b, c\}$ to $\{1, 2, 3, 4\}$?*

(c) *Is it possible to construct a 1-to-1 correspondence from one finite set to another if the two sets have different numbers of elements?*

Chapter 2

Combinatorics

"All right," said Benjy. *"What do you get if you multiply six by seven?"*

"No, no, too literal, too factual," said Frankie, "wouldn't sustain the punters' interest."

Again they thought.

Then Frankie said, "Here's a thought. *How many roads must a man walk down?"*

"Ah," said Benjy. "Aha, now that does sound promising!" He rolled the phrase around a little. "Yes," he said, "that's excellent! Sounds very significant without actually tying you down to meaning anything at all. *How many roads must a man walk down? Forty-two.* Excellent, excellent, that'll fox 'em. Frankie, baby, we are made!"

Douglas Adams

2.1 Finite sets

A set is said to be *finite* if it has finitely many elements. If a set X is finite, the number of elements in it is denoted by $|X|$ and is called, *the size of X*.

23

Example 2.1 $|\{a, b, c, \ldots, z\}| = 26;$ $|\{\clubsuit, \heartsuit, \diamondsuit, \spadesuit\}| = 4;$ $|\emptyset| = 0$

2.2   The Rule of Product and The Rule of Sum

If X and Y are finite sets, then

$$|X \times Y| = |X||Y| \tag{2.1}$$

Equality (2.1) says that a table with $|X|$ rows and $|Y|$ columns has $|X||Y|$ cells.

Example 2.2 *In a spreadsheet, let $X = \{1, 2, 3, 4\}$ be the set of the top 4 rows, and $Y = \{A, B, C\}$ be the set of the leftmost 3 columns. Counting all the cells that are both in rows 1 to 4 and in columns A to C, one finds exactly $3 \cdot 4 = 12$ such cells. Conducting similar experiments with other choices of the sets X, Y (and drawing the grid-like diagram suggested by the appearance of a spreadsheet), one can convince oneself that (2.1) holds.*

If X and Y are disjoint (section 1.4), then

$$|X \cup Y| = |X| + |Y| \tag{2.2}$$

Equality (2.2) says that if two lists with no item in common are joined into one list, the resulting number of items is the sum of those on the original two lists.

In many texts, (2.1) is called the *Rule of Product*; (2.2), the *Rule of Sum.*

Exercise 2.1 *Experimentally test the following statement: If $X_1, \ldots, X_n$ are finite sets, then*

$$|X_1 \times \ldots \times X_n| = |X_1| \cdot \ldots \cdot |X_n| \tag{2.3}$$

Exercise 2.2   (a) *If X and Y are finite sets with at least one element in common, then formula (2.2) fails. Why? What is a correct formula?*

(b) *How many integers between 1 and 480, inclusive, are divisible by either 2 or 3 or both?*

(c) *How many integers ≥ 1 and ≤ 480 do not have a common factor with 12? [First find how many such integers do have a common factor with 12.]*

2.3 The Factorial Function

Let n be an integer ≥ 0. The *factorial* of n is the positive integer equal to

$$\begin{cases} 1 & if \quad n = 0 \\ 1 \cdot 2 \cdot \ldots \cdot n & if \quad n > 0 \end{cases}$$

The factorial of n is denoted $n!$.

The observation

$$n! = \underbrace{1 \cdot 2 \cdot \ldots \cdot (n-2) \cdot (n-1)}_{(n-1)!} \cdot n$$

leads to the recurrence

$$n! = (n-1)! \, n \quad \text{for } n = 1, 2, 3, \ldots \tag{2.4}$$

Exercise 2.3 *Compute the factorials $0!, 1!, 2!, 3!, 4!,$ and $5!$. [Use of (2.4) will save work.]*

Exercise 2.4 *Express the product $9 \cdot 10 \cdot \ldots \cdot 126 \cdot 127$ in terms of factorials.*

2.4 Counting functions from one finite set to another

Theorem 2.3 *If X and Y are finite sets, then the total number of functions from X to Y is $|Y|^{|X|}$.*

Proof. Let n be the number of elements in X, so we can write

$$X = \{x_1, x_2, \ldots, x_n\}$$

Let f be a function from X to Y. By writing out all values of f, we obtain the ordered n-tuple

$$(f(x_1), f(x_2), \ldots, f(x_n))$$

of element of Y. By (2.3), there are $|Y|^n = |Y|^{|X|}$ such n-tuples. Conversely, every such n-tuple gives a function from X to Y. This completes the proof.

Exercise 2.5 *A palette of 8 different colors is avaiable to color the faces of a cube. Each side must painted one color; painting multiple (or even all) faces the same color is allowed. How many colorings are there? [Each coloring is a function from the set of faces to the set of colors.]*

Exercise 2.6   (a) *Let $X = \{a, b\}$. Determine how many subsets X has by writing out all of them (or their membership functions).*

(b) *Let $X = \{a, b, c\}$. Determine how many subsets X has by writing out all of them (or their membership functions).*

(c) *Suppose X is a finite set. Denote by 2^X the set of all subsets of X. (2^X is called, the power set of X.) Calculate $|2^X|$ in terms of $|X|$.*

An intermediate task in many combinatorial problems is to count all the different ways of forming a sequence of a given length k by choosing k distinct elements from a given set and arranging them in some order. The following theorem tells how.

Theorem 2.4 *If X is a finite set with n elements, then the number of sequences consisting of k distinct elements of X is*

$$\frac{n!}{(n-k)!} \tag{2.5}$$

Proof. A procedure for constructing such a sequence, which we can denote

$$x_1, x_2, \ldots, x_k, \tag{2.6}$$

is simply to *choose* each of the elements (2.6) from the set X so that no two are the same. For x_1, any one of the n elements of X can be chosen. For x_2, the number of choices is $(n-1)$, as one element of X has already been used for x_1. For x_3, only $(n-2)$ choices are left, and so on. Finally, for x_k there remain $(n-k+1)$ choices. Hence, by the Rule of Product (section 2.2) the number of ways of constructing the sequence is

$$n \cdot (n-1) \cdot (n-2) \cdot \ldots \cdot (n-k+1),$$

which is precisely the quantity (2.5). This completes the proof.

Exercise 2.7 *How many ways are there to form words consisting of 4 distinct letters from the lowercase English alphabet $\{a, b, \ldots, z\}$?*

Exercise 2.8 *A password must consist of five distinct characters, each character allowed to be either a lowercase English letter or one of the symbols*

$$!, \#, @, \text{-}$$

How many such passwords are possible?

2.5 Permutations

Given a deck of cards, a shuffling of the deck can be described without referring to any specific order in which the cards were originally stacked[1]. How? By instructing one to shuffle the deck so that, for example:

1. The 6 of clubs now occupies the position in the deck previously occupied by the 10 of spades;

2. The king of diamonds now occupies the position previously occupied by the queen of hearts;

3. The jack of clubs occupies the same position as previously;

---

[1]This crucial fact is often obscured by confusing terminology such as "order matters" and "order does not matter." Avoid the framework of such terminology and rely on the clear definitions instead.

4. The queen of hearts now occupies the position previously occupied by the 9 of spades;

5. The ace of spades now occupies the position in the deck previously occupied by the 7 of clubs;

etc. (for a total of 52 such instructions).

Thus, a shuffling of the deck is, essentially, a 1-to-1 correspondence (section 1.8) from the set of the 52 cards to itself.

A 1-to-1 correspondence from a finite set to itself is called, *a permutation* of that set. Given a finite set

$$X = \{x_1, x_2, \ldots, x_n\},$$

a permutation f of X can be described by the table

x_1	x_2	$\ldots$	x_N
$f(x_1)$	$f(x_2)$	$\ldots$	$f(x_n)$

Example 2.5 *The permutation*

$$f(\clubsuit) = \spadesuit, \ f(\heartsuit) = \diamondsuit, \ f(\diamondsuit) = \clubsuit, \ f(\spadesuit) = \heartsuit$$

of the set $X = \{\clubsuit, \heartsuit, \diamondsuit, \spadesuit\}$, *is described by the table*

$\clubsuit$	$\heartsuit$	$\diamondsuit$	$\spadesuit$
$\spadesuit$	$\diamondsuit$	$\clubsuit$	$\heartsuit$

The analogy between a permutation and a shuffling of a deck of cards, given at the beginning of this section, leads to another way of depicting the permutation f; namely, by listing the elements of X, and drawing an arrow from each element x to $f(x)$.

Example 2.6 *The permutation defined in example 2.5 can be pictured thus:*

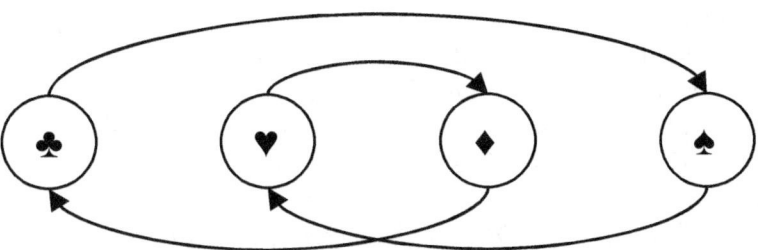

The same permutation can be pictured thus:

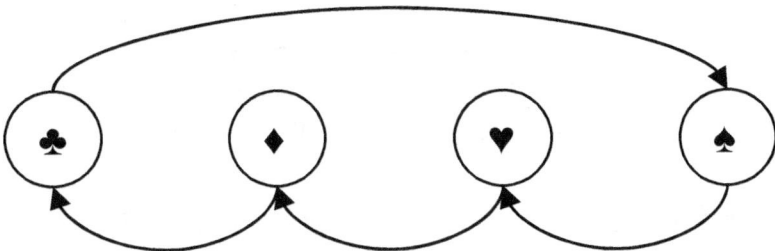

The next example demonstrates the geometry of permutations.

Example 2.7 *If the vertices of a square are labeled A, B, C, D clockwise, the permutation that maps A to B, B to C, C to D, and D to A is effected by a clockwise 90-degree rotation of the square about its center. Furthermore, if all the labels are deleted, then the rotated square is indistinguishable from the original one. The rotation is thus a one-to-one correspondence from the plane to itself that maps the square to itself and preserves distances between points of the square. Such a one-to-one correspondence is called a* symmetry *of the square. Thus, the latter permutation of the square's vertices arises from a symmetry of the square. It is a worthwhile exercise to find the symmetries of the square that give rise to*

- *the permutation that interchanges the vertices A and B, and interchanges the vertices C and D;*

- *the permutation that interchanges the vertices B and D.*

*By drawing the square and labeling the vertices as described, and
then imagining a mirror perpendicular to the plane of the paper and
covering the diagonal AC of the square, one sees that the latter symmetry (interchanging B and D) is a reflection through that mirror.
This reflection is called,* a reflection through the axis AC, *and is an
example of the* axial symmetry *covered in elementary geometry (see
[7]).*

*Among other examples of polytopes whose symmetries give rise to
permutations of vertices are an equilateral triangle, a regular octagon,
a cube, and a regular tetrahedron. Not every permutation comes from
a symmetry of a polytope; e.g., if you glue one face of a cube to
the table, and rotate the opposite face in the horizontal plane by 90
degrees (thus permuting the vertices of the top face), the cube's shape
is broken, i.e. this is not a symmetry of the cube.*

*For a surprising fundamental connection between symmetries and
roots of polynomials, see [2] (a clear and concise exposition addressed
to highschoolers).*

Theorem 2.8 *A set with n elements has $n!$ permutations.*

Proof. Write this set as $X = \{x_1, x_2, \ldots, x_n\}$. To specify a permutation f of X is to choose distinct values for all of $f(x_1), f(x_2), \ldots, f(x_n)$.
These values are to be elements of X. We proceed as follows. Choose
$f(x_1)$ (n choices). Once $f(x_1)$ is chosen, $(n-1)$ choices are available for $f(x_2)$. Thus, the number of ways to choose $f(x_1)$ and $f(x_2)$
simultaneously is $n(n-1)$. Once $f(x_2)$ is chosen, $(n-2)$ choices
remain for $f(x_3)$, so the number of ways to choose simultaneously
$f(x_1)$, $f(x_2)$, and $f(x_3)$ is $n(n-1)(n-2)$. Continuing this way, we
see that there are $n(n-1) \cdot \ldots \cdot 2 \cdot 1 = n!$ ways of choosing f. This
completes the proof.

2.6   Counting rearrangements of sequences

One often needs to count the number of ways to rearrange a string
of symbols, originally given in some order,–that is, to rearrange the
terms in a sequence. In this section, we learn to make such counts.
The only sequences considered here are words, i.e. sequences of English letters (not necessarily with a meaning). By the length of a

word, we shall mean the number of letter occurrences in the word; e.g., the word $MASSACHUSSETTS$ has length 14. This usage of the term *length* clarifies the confusion sometimes caused by the phrase "rearrange the letters in a word." What one is really rearranging are *occurrences* of letters, with a letter allowed to occur more than once.

If a word consists of N distinct letters, then, by theorem 2.8, the number of ways to rearrange it is $N!$. Notice that, in this text, the original word is also regarded as a rearrangement of itself; this usage saves verbiage.

Example 2.9 *The $3!$ rearrangements of the word CAT are:*

$$CAT, ACT, TAC, TCA, CTA, ATC$$

The problem becomes more difficult if a word contains multiple occurrences of the same letter. For example, consider the word $BALL$. In any rearrangement of this word, an interchange of the two occurrences of L produces no new rearrangement. Thus, there are fewer than $4!$ rearrangements, and we now describe, how many exactly.

2.6.1 One letter occurring twice, the others once each

Starting with a simple example, consider again the word $BALL$. The difficulty, as mentioned above, is that the occurrences of L are indistinguishable. To resolve this, we temporarily index these occurrences with subscripts, so as to distinguish between them. Each subscript indicates the place of the occurrence in the word. The letter L occurs in the word $BALL$ in the 3rd and 4th places. This leads to the interim problem of counting all the rearrangements of the word

$$BAL_3L_4,$$

of length 4, treating L_3 and L_4 as distinct letters. We now group the rearrangements of this word so that, within each group, removing the subscripts makes all the rearrangements the same. Writing all rearrangements in a table, where each column is such a group, we obtain

$$
\begin{array}{cccc}
BAL_3L_4 & ABL_3L_4 & AL_3BL_4 & \cdot\ \cdot\ \cdot \\
BAL_4L_3 & ABL_4L_3 & AL_4BL_3 & \cdot\ \cdot\ \cdot
\end{array}
$$

Thus, the number of ways to rearrange the word $BALL$ is the number of columns in the latter table. By theorem 2.8, the table has 4! entries. Each column has $2! = 2$ entries, so there are $4!/2 = 12$ columns.

2.6.2   One letter occurring three times, the others once each

Consider the word $STRESS$. Here, the letter S occurs three times; all the other letters, once each. Using the same approach as in section 2.6.1, we distinguish the occurrences of S by subscripts. The letter S occurs in the 1st, 5th, and 6th places. We obtain

$$S_1TRES_5S_6 \tag{2.7}$$

We now consider all 6! rearrangements of this word, grouping them as follows. Two rearrangements will be placed in the same group if, on removing the subscripts, they become indistinguishable. Tabulating these rearrangements of the word (2.7) with each group occupying one column, we obtain the table

$$
\begin{array}{llll}
S_1TRES_5S_6 & RS_1TES_5S_6 & S_1RS_5TS_6E & \cdot\ \cdot\ \cdot \\
S_1TRES_6S_5 & RS_1TES_6S_5 & S_1RS_6TS_5E & \cdot\ \cdot\ \cdot \\
S_6TRES_1S_5 & RS_6TES_1S_5 & S_6RS_1TS_5E & \cdot\ \cdot\ \cdot \\
S_6TRES_5S_1 & RS_6TES_5S_1 & S_6RS_5TS_1E & \cdot\ \cdot\ \cdot \\
S_5TRES_1S_6 & RS_5TES_1S_6 & S_5RS_1TS_6E & \cdot\ \cdot\ \cdot \\
S_5TRES_6S_1 & RS_5TES_6S_1 & S_5RS_6TS_1E & \cdot\ \cdot\ \cdot \\
\end{array}
$$

In the latter table, two rearragengements in the same column differ only by a permutation of the three subscripts $1, 5, 6$. Thus, each column has 3! entries. Consequently, there are $6!/3! = 120$ columns, and this is how many ways there are to rearrange the word $STRESS$.

2.6.3   More than one letter occurring more than once

Consider the word $ACACIAS$. Here, two of the letters, A and C, have multiple occurrences. We first index only the occurrences of A:

$$A_1CA_3CIA_6S \tag{2.8}$$

Using the approach in section 2.6.2, one finds $7!/3!$ ways to rearrange the word (2.8). Grouping these rearrangements so that two in a group differ only by a permutation of the C's with subscripts (with the occurrences of all the other letters staying in their positions) we obtain that the word $ACACIAS$ has a total of

$$(7!/3!)/2! = \frac{7!}{3!2!}$$

rearrangements.

2.6.4 The general formula

Considering words with three or more letters having multiple occurrences, one can apply the above approach to arrive at the general formula, given in the following theorem.

Theorem 2.10 *Given a word where the numbers of occurrences of each letter are, respectively,*

$$k_1, k_2, \ldots, k_m$$

(and, consequently, with the word having length $k_1 + k_2 + \ldots + k_m$), there are

$$\frac{(k_1 + k_2 + \ldots + k_m)!}{k_1! \cdot k_2! \cdot \ldots \cdot k_m!}$$

ways to rearrange this word.

Exercise 2.9 *A robot is capable of executing a sequence commands, where each command is either W (for "take a step Westward") or N ("take a step Northward"). All robot steps have the same length.*

Points A and B are on the diagonally opposite corners of a rectangle. Two of its sides are oriented North-to-South; the other two, East-to-West. The rectangle is 27 robot steps long and 16 robot steps wide. How many different command sequences will accomplish the task of navigating the robot from A to B?

2.7   Counting subsets of a finite set

In combinatorics and its applications, one must often determine, *how many ways there are to pick a subset of k objects from a set of n.* This problem can be restated in a much less abstract form; namely, *how many ways are there to rearrange the word*

$$AA \ldots ABB \ldots B,$$

where A occurs k times, and B occurs $(n - k)$ *times?*   In this latter form, the question is immediately answered by theorem 2.10 on taking $k_1 = k$ and $k_2 = n - k$: there are

$$\frac{(k_1 + k_2)!}{k_1!\, k_2!}$$

rearrangements. This answers the original question, i.e. the one in the opening sentence of this section, as follows.

Theorem 2.11 *Let n, k be integers satisfying $n \geq k \geq 0$, and let X be a finite set with n elements. Then X has precisely*

$$\frac{n!}{(n - k)!\, k!} \tag{2.9}$$

subsets of cardinality k.

The quantity (2.9) is denoted C_k^n or $\begin{pmatrix} n \\ k \end{pmatrix}$ (read "n choose k," since this is the number of ways to choose a set of k objects out of n).

Exercise 2.10 *A homeowner has 100 Haloween candy bars, no two bars being the same. A visitor asks for 6 candy bars. In how many ways can the request be granted if*

   (a) *the visitor is happy with any 6 of the 100 bars?*

   (b) *the visitor wants a Sneaker to be among the 6 requested bars?*

   (c) *the visitor wants a Sneaker and a Milky Way to be among the 6 requested bars?*

Exercise 2.11 *(Partitions of Integers.)*

   (a) *How many ways are there to express the number 10 as a sum*

$$a_1 + a_2 + a_3 + a_4 = 10$$

of strictly positive integers a_1, a_2, a_3, a_4? Such a sum is called a partition *of 10. (In this exercise, permuting the summands is considered to give a different partition; e.g., the partitions*

$$a_1 = 3, a_2 = 3, a_3 = 2, a_4 = 2$$

and

$$a_1 = 3, a_2 = 2, a_3 = 3, a_4 = 2$$

are considered distinct. Determining how to count partitions that differ by a permutation of the summands as the same is left to the reader.)

   (b) *As in the previous part of the exercise, but this time allowing the integers a_k to be nonnegative. [If $a_k \geq 0$, then $a_k + 1 > 0$. Now take $b_k = a_k + 1$ and note that we must have*

$$
\begin{aligned}
& b_1 + b_2 + b_3 + b_4 \\
= \ & (a_1 + 1) + (a_2 + 1) + (a_3 + 1) + (a_4 + 1) \\
= \ & a_1 + a_2 + a_3 + a_4 + 4 \\
= \ & 14
\end{aligned}
$$

This reduces the problem to the previous case.]

   (c) *How many ways are there to distribute n indistinguishable marbles in k numbered jars? (The marbles are indistinguishable in the sense that swapping two marbles from two distinct jars does* not *give a new distribution.)*

   (d) *Same question with the additional requirement that*

1. *no jar be empty [First, put a marble in each jar. Then, count the ways of distributing the remaining marbles.]*

2. *at least one jar be empty*

3. *precisely one jar be empty*

4. *jar #1 contains an even number of marbles*

2.8   Pascal's Identity

Suppose a set X has n elements, and suppose k is an integer such that $0 \leq k \leq n$. By theorem 2.11, the set X has C_k^n subsets of size k.

On the other hand, consider the following procedure for choosing from X a subset with k elements.

1. Choose a specific element x of X. Let Y denote the set consisting of the remaining elements of X (i.e., $Y = X \setminus \{x\}$). The subsets of X that have k elements are of two types: those that do contain x (call them, *type I*), and those that do not (*type II*).

2. To count the subsets of type I is to count the subsets of Y of size $(k-1)$; by theorem 2.11, there are C_{k-1}^{n-1} of them.

3. To count the subsets of type II is to count the subsets of Y of size k; by theorem 2.11, there are C_k^{n-1} of them.

The above procedure shows that the total number of ways to choose from X a subset with k elements (by theorem 2.11, this number is C_k^n) equals $C_{k-1}^{n-1} + C_k^{n-1}$. We have obtained the equality

$$C_k^n = C_{k-1}^{n-1} + C_k^{n-1}, \tag{2.10}$$

called *Pascal's Identity*[2].

---

[2]Blaise Pascal (1623 - 1662) was a French mathematician, physicist, and philosopher.

Exercise 2.12   (a) *Consider the following three lines of numbers:*

$$1$$
$$1 \quad 1 \tag{2.11}$$
$$1 \quad 2 \quad 1$$

Observe that these lines are precisely

$$C_0^0,$$

$$C_0^1 \; C_1^1,$$

and

$$C_0^2 \; C_1^2 \; C_2^2$$

Arranged as in (2.11), they form the top three rows of a "pyramid" whose n-th row (counting top down) is

$$C_0^n, \; C_1^n, \; \ldots, \; C_n^n$$

The constructed pyramid of numbers is called Pascal's Triangle *(although it appears in records of mathematical works from Greece, India, Persia, China, and Italy earlier than Pascal's lifetime). Use Pascal's Identity (2.10) to calculate the next several rows of the pyramid. [If difficulties arise, skip to the last part of the exercise.]*

(b) *Verify that $C_k^n = C_{n-k}^n$. This observation implies a certain symmetry for Pascal's Triangle. Try to describe that symmetry.*

(c) *By flushing all the rows of Pascal's Triangle to the left, one obtains the layout*

$$1$$
$$1 \quad 1$$
$$1 \quad 2 \quad 1$$

$$\cdot \qquad \qquad \cdot$$
$$\cdot \qquad \quad \cdot$$
$$\cdot \qquad \qquad \cdot$$

*This layout is more convenient for computing the entries
of Pascal's Triangle using a spreadsheet. Compute the
first 10 rows of the Triangle.*

(d) *Read appendix C and expand each of the following:*

- $(x + y)^2$
- $(x + y)^3$
- $(x + y)^4$

(e) *Notice that*

$$
\begin{aligned}
11^0 &= 1 \\
11^1 &= 11 \\
11^2 &= 121 \\
11^3 &= 1331
\end{aligned}
$$

*This seems to suggest the pattern that the digits of 11^n
form the n-th row of Pascal's Triangle. Use a spreadsheet
to determine how many rows of the Triangle follow the
pattern. Why does it break down?*

2.9   A survey of further combinatorics

If you need further knowledge of combinatorics (whether for upcoming courses or exams, work, business, or personal interest), [9] is a classical exposition and a source of exercises, and is the first volume of four. These volumes introduce the reader to *algorithms*. An algorithm is, loosely speaking, a sequence of instructions that can be carried out by a computer. (Strictly speaking, at a more advanced level, an algorithm can be thought of as a computational model, such as a *Turing machine* [10].) The following exercise asks for a simple algorithm.

Exercise 2.13 *A farmer must take a wolf, a goat, and a cabbage across a river in a boat. The boat will fit only one of these three objects (besides the farmer, who must do the rowing). Without the farmer present, the wolf will eat the goat, and the goat (if alive) will eat the cabbage. How should the farmer proceed?*

A subject that substantially overlaps with combinatorics (and algorithm theory) is *graph theory*. The concept of a graph arises in numerous applications in engineering: transportation, networks, scheduling, city planning, to name a few. A clear elementary (and example-oriented) introduction to graph theory is found in [13]; the same author also has written books on more advanced graph theory.

Chapter 3

Elementary probability theory

> ''...Your arrival on the planet has caused considerable
> excitement. It has already been hailed, so I gather,
> as the third most improbable even in the history
> of the Universe.''
>
> ''What were the first two?''
>
> ''Oh, probably just coincidences''...

Douglas Adams

In practice, a formula is only as good as its predictions are accurate. (For example, Newton's laws do well to predict motion of large bodies moving slowly, but do much worse for elementary particles moving nearly at the speed of light.) Every prediction, in turn, requires that some information be given to start with. In fact, to predict is *to infer new information from that given.*

To give a good prediction, one must generally have detailed insight into the process whose outcome is being predicted. In most processes, however, such insight is lacking. This makes the outcome of an experiment impossible to predict with certainty. (A few examples are stock market behavior, social trends, quantum-mechanical phenomena, the trajectory of a ship being moored to the quay, or that of an airplane flying in changing wind.) Yet, such predictions are badly needed: business strategies must be devised, ships moored, and airplanes navigated. What can one do?

3.1   Sample spaces, trials, and events

The compromise offered by probability theory is as follows. Instead of aiming to predict the outcome, one aims to predict the chance that the outcome will lie in a given set. In detail, one begins by specifying the set of *all* possible outcomes; this set is called, *the sample space* of the experiment. (The act of carrying out the experiment once is called, *a trial*.) For every subset of the sample space, one tries to measure quantitatively the chance that the outcome will be in that subset. The subsets of the sample space are called, *events*. This term is used since the description of an event often comes in prose.

Exercise 3.1 *Given events E and F in a sample space S, express the following events in terms of set operations (section 1.4).*

(a) *Both E and F have occurred.*

(b) *At least one of E, F has occurred.*

(c) *Event E has occurred, while event F has not.*

(d) *Event E has not occurred.*

Example 3.1 *Consider the experiment consisting of tossing a coin twice in a row. The sample space is*

$$\{TT, TH, HT, HH\}$$

Four of the events are:

- *The event "The second toss yields Tails" is the subset $\{TT, HT\}$ of the sample space.*

- *The event "Neither toss yields Heads" is the subset $\{TT\}$, consisting of the one outcome TT.*

- *The event "The first toss yields both Heads and Tails" is the subset $\emptyset$ of the sample space.*

- *The event "The second toss yields either Heads or Tails" is the subset equal to the entire sample space.*

The latter example illustrates, in particular, why in every experiment the event $\emptyset$ is called, *the impossible event*, while the event equal to the entire sample space is called, *the certain event*.

Two events that are disjoint as sets (see section 1.4) have no outcome in common, hence are called *mutually exclusive*.

3.2 Probability measures

The quantitative measurement of the chance of an event, mentioned at the beginning of section 3.1, aims to assign to each event a number that somehow reflects "how likely" that event is to occur. This number is called, *the probability of the event*. One usually constructs probabilities using either the *frequentist approach* (used in this book) or the *Bayesian*[1] *approach*. Whichever approach is taken, one always makes the following assumptions.

Assumption 3.1   (a) *It is in principle possible to conduct as many trials as one wishes (even infinitely many).*

   (b) *The sample space of the experiment is specified prior to all trials and remains unchanged from trial to trial.*

   (c) *Different trials of the experiment can yield different outcomes, which generally cannot be predicted with certainty.*

   (d) *The nature of the process explored in the experiment remains unchanged from trial to trial. (Consequently, the chance that a given event will result, remains the same from trial to trial.)*

The frequentist construction of probabilities is as follows. Suppose E is an event. Conducting N trials of the experiment, each

---

[1]Rev. Thomas Bayes (1702 - 1761) was a British clergyman and mathematician.

yielding an outcome, one asks, How many of the N outcomes lie in E? Dividing the number of such outcomes by N, we obtain a ratio, called *the relative frequency of E* observed from the N trials.

Example 3.2 *Consider the experiment consists in tossing a coin once, and let E be the event "Tails." Suppose $N = 20$ trials have yielded the outcome sequence*

$$H, T, H, T, T, T, H, H, T, T, T, H, H, T, T, H, T, T, H, H$$

Of these, 11 have resulted in the event E, whose relative frequency is therefore

$$\frac{11}{N} = \frac{11}{20} \tag{3.1}$$

Exercise 3.2 *Consider the experiment described in example 3.1. For $N = 5$, and then for $N = 10$, conduct N trials and calculate the relative frequency of the event "Only one Heads appears in the outcome." (Or, if possible, write a computer program to carry out the above trials and calculations, and see what relative frequencies it gives for larger values of N, e.g. for $N = 200, 500, 10000$.)*

For a given N outcomes, let $\phi_N(E)$ denote the relative frequency of event E. By doing simple experiments with small N, one can establish the following facts:

- the relative frequency of any event E cannot be negative or exceed 1:

$$0 \le \phi_N(E) \le 1; \tag{3.2}$$

- the relative frequency of the impossible event is 0:

$$\phi_N(\emptyset) = 0; \tag{3.3}$$

- the relative frequency of the certain event S is 1:

$$\phi_N(S) = 1; \tag{3.4}$$

- If events E and F are mutually exclusive, then

$$\phi_N(E \cup F) = \phi_N(E) + \phi_N(F) \tag{3.5}$$

While different sequences of N trials generally give different sequences of outcomes, assumptions 3.1 suggest that, for large N, the relative frequency of a given event E should come out approximately the same in each sequence of N trials. Furthermore, as N tends to infinity, the relative frequencies should tend to an *exactly the same value*. This common value of the relative frequencies obtained for E is called, *the probability of E*. A function that assigns to each event its probability is called, *a probability measure*.

The rigorous axiomatic construction of probabilities, omitted here, involves taking a limit of the relative frequencies as N tends to infinity. Limit theory, in turn, says that the properties stated in equations (3.2), (3.3), (3.4), and (3.5) for relative frequencies are "inherited" by probabilities. Namely, if S denotes the sample space, and $\mathbf{P}(E)$ the probability of event E, one has

$$0 \leq \mathbf{P}(E) \leq 1 \quad \text{for all events } E \qquad \text{(a)}$$

$$\mathbf{P}(S) = 1 \qquad \text{(b)}$$

$$\mathbf{P}(E \cup F) = \mathbf{P}(E) + \mathbf{P}(F) \quad \text{with } E, F \text{ mutually excl.} \quad \text{(c)}$$

$$(3.6)$$

(In texts that follow the axiomatic approach of Kolmogorov[2], e.g. [8], the three properties (3.6) are taken as the *definition* of a probability measure s**P**.) From (3.6.b) and (3.6.c), one calculates

$$1 = \mathbf{P}(S) = \mathbf{P}(\emptyset \cup S) = \mathbf{P}(\emptyset) + \mathbf{P}(S) = \mathbf{P}(\emptyset) + 1$$

This calculation (in which the reader should justify every equality) yields

$$\mathbf{P}(\emptyset) = 0$$

The latter equality (which corresponds to experimental fact (3.3) about relative frequencies) shows that the impossible event $\emptyset$ has probability zero. The following example shows that a nonempty (i.e., possible) event can also have probability zero.

Example 3.3 *Consider the experiment of tossing a dart at a solid square. The sample space, therefore, consists of all the points on*

---

[2]A. N. Kolmogorov (1903 - 1987) was a Russian and Soviet mathematician.

or inside the boundary of the square. Let the probability measure **P** *assign to each event E the geometric area of E^3. An event consisting of one point is nonempty, yet has probability zero.*

Example 3.4 *A probability measure for the experiment of tossing a coin with a 2-to-1 bias toward Tails is specified by the following table:*

x	T	H
$\mathbf{P}(x)$	2/3	1/3

Exercise 3.3 *This exercise consists in using a Google Documents spreadsheet to simulate a coin with a given bias. For convenience, use the number 0 to denote T, and the number 1 to denote H. Simulate fifteen tosses of a coin biased in each of the following ways.*

(a) $\mathbf{P}(0) = 0.2$, $\mathbf{P}(1) = 0.8$

(b) $\mathbf{P}(0) = 2/3$, $\mathbf{P}(1) = 1/3$

(c) $\mathbf{P}(0) = 0.6$, $\mathbf{P}(1) = 0.4$

An event consisting of a single outcome is called, *an elementary event.* While the probability of an elementary event $\{x\}$ (with x the corresponding single outcome) is properly written $\mathbf{P}(\{x\})$, the curly brackets are often omitted, resulting in the notation $\mathbf{P}(x)$.

Property (3.6c) shows that, if an event E consists of finitely many outcomes, i.e.

$$E = \{x_1, x_2, \ldots, x_M\},$$

then

$$\mathbf{P}(E) = \mathbf{P}(x_1) + \mathbf{P}(x_2) + \ldots + \mathbf{P}(x_M)$$

Thus, to specify a probability measure on a finite sample space S, it suffices to specify the probabilities of the elementary events.

Exercise 3.4 *Describe a sample space and two different probability measures for each of the following experiments.*

---

[3]Not every subset of the square has a well-defined geometric area, but this technicality is beyond the scope of the book and will not hinder this exposition. The advanced subject addressing this issue is *measure theory*; see section 3.6 for a brief discussion and indication of further literature.

(a) *Tossing a coin twice. [Two different probability mea-sures can arise, for example, depending on whether the coin is fair or biased.]*

(b) *Tossing a die once.*

(c) *Tossing a die twice.*

(d) *Tossing a coin three times.*

(e) *Tossing a dart at a partially blackened square, shown in the figure below.*

Exercise 3.5 *If E is an event in a sample space S, then the* comple-ment *(or* complementary event*) of E is the set-theoretic complement*

$$E^c = S \setminus E$$

(also read "The event 'not E'.") For each of the following statements, determine whether it is generally true or false:

(a) $\mathbf{P}(E^c) = 1 - \mathbf{P}(E)$

(b) $\mathbf{P}(F) = \mathbf{P}(F \cap E) + \mathbf{P}(F \cap E^c)$

(c) *If an event F is such that $E \subset F$, then $\mathbf{P}(E) \leq \mathbf{P}(F)$.*

(d) *If an event F is such that $E \subset F$, then $\mathbf{P}(E) < \mathbf{P}(F)$.*

[If having difficulty, examine the analogous assertions for relative frequencies.]

Exercise 3.6 *(Buffon's needle problem.)   A needle 1 cm long is tossed onto a table covered with parallel lines 1 cm apart.   What is the probability that the needle will cross one of the lines?*

Exercise 3.7 *(Bernoulli trials[4].) Suppose a coin is such that a toss yields* Heads *with probability p, and* Tails *with probability $1-p$. Consider the experiment of tossing the coin n times.   What is the probability that k of the n outcomes will be* Tails*? Hint: Use theorem 2.11.*

3.3   Conditional probability

A demand for a probabilistic prediction often comes with some prior information or stipulation about the outcome.   Examples include needing to know the probability that

- a supermarket customer who has bought diapers will also buy beer;

- a patient who has a disease will test for it positively;

- a patient who does not have a disease will test for it negatively;

- a convicted burglar will burgle again within 3 years of release from jail;

- an inkjet printer that malfunctions does so because of a deficient mechanical gear.

All these probabilities are of the form "the probability of an event A given that an event B has occurred." They are known as *conditional probabilities* (an explanation is given below) and play a crucial role in designing various decision strategies, whether in business, manufacturing, medicine, or another field.

---

[4]Jacob Bernoulli (also known as James or Jacques) (1654 - 1705) was one of the prominent mathematicians of the Bernoulli family

To understand conditional probability, consider events A and B in a sample space S with a probability measure $\mathbf{P}$. Suppose some N trials have been carried out and, among the resulting N outcomes, the events B and $(A \cap B)$ have turned up, respectively, N_B times and N_{AB} times. To study how the occurrence of B affects that of A, we ask, *What fraction of the N_B outcomes in B is also in A?* This fraction is

$$\frac{N_{AB}}{N_B} \tag{3.7}$$

If B was the entire sample space, the fraction (3.7) would give the relative frequency of event $(A \cap B)$ in the above N outcomes, and so can be thought to approximate *the probability of A given the information that B has occurred.* Dividing the numerator and the denominator in (3.7) by N, we obtain the fraction

$$\frac{N_{AB}/N}{N_B/N},$$

whose value is the same as that of (3.7), but whose form is

$$\frac{\text{relative frequency of } (A \cap B)}{\text{relative frequency of } B}$$

Finally, replacing each relative frequency by the probability it approximates, we obtain the ratio

$$\frac{\mathbf{P}(A \cap B)}{\mathbf{P}(B)},$$

called *the conditional probability of A given B*, and denoted by $\mathbf{P}(A|B)$; thus, the definition of conditional probability can be written

$$\mathbf{P}(A|B) = \frac{\mathbf{P}(A \cap B)}{\mathbf{P}(B)}, \tag{3.8}$$

Multiplying both sides of (3.8) by $\mathbf{P}(B)$, one obtains *Bayes's formula*

$$\mathbf{P}(A \cap B) = \mathbf{P}(B)\mathbf{P}(A|B), \tag{3.9}$$

useful (despite its simplicity) when one must calculate $\mathbf{P}(A \cap B)$ when given only $\mathbf{P}(B)$ and $\mathbf{P}(A|B)$. An application where such a calculation is needed is described, below, in example 3.5.

The difference between the probability $\mathbf{P}(A)$ and the conditional probability $\mathbf{P}(A|B)$ measures how dependent A is on B. If the two probabilities are equal, then the occurrence of B has no bearing on the probability of A. Hence the following concept: events A and B are said to be *statistically independent* if

$$\mathbf{P}(A|B) = \mathbf{P}(A),$$

where $\mathbf{P}(B)$ is nonzero.

Exercise 3.8 *In the experiment of drawing, randomly, a card from a deck of* 52 *cards, let A be the event that a queen is drawn, B the event that a spades card is drawn, C the event that a black suit card (spades or clubs) is drawn, and D the event that a red suit is drawn.*

(a) *What are the conditional probabilities $\mathbf{P}(A|B)$ and $\mathbf{P}(B|A)$?*

(b) *Are the events A, C independent?*

(c) *Are the events A, D independent?*

(d) *Are the events B, D independent?*

Conditional probabilities are an essential instrument in *statistical inference*. The remainder of this section is a brief introduction to the central questions of statistical inference. A detailed exposition of the topic is beyond the scope of this book (but can be found in [16] and in [11]).

In practice, one sometimes has a collection of events which constitute certain *evidence*, and needs to compute the probability of an event called a *hypothesis*. The only knowledge available are the individual probabilities of the evidence events, and the conditional probability of the hypothesis on each evidence event.

Example 3.5 *A test of a patient's blood for a bacterial infection can yield one of two possible outcomes, $\oplus$ (positive, i.e. the blood is infected) and $\ominus$ (negative). From the engineering specifications of the test equipment, one knows the probabilities of a false positive and a false negative. From clinical data, one knows the probabilities $\mathbf{P}(\oplus)$ and $\mathbf{P}(\ominus)$.*

If I and C are, respectively, the events "actually infected" and "actually clean," the probability of a false positive is the conditional probability $\mathbf{P}(\oplus|C)$. A patient who has tested positive, however, wants to know, not the latter probabilities, but the probability that she or he is actually infected: $\mathbf{P}(I)$. Therefore, it is desirable to compute the unknown probability $\mathbf{P}(I)$ from the known ones. Since the events $\oplus$ and $\ominus$ are mutually exclusive, and since $\oplus \cup \ominus$ is the entire sample space S, we have

$$I = (I \cap \oplus) \cup (I \cap \ominus),$$

where $(I \cap \oplus)$ and $(I \cap \ominus)$ are mutually exclusive. By a solution to exercise 3.5(2), we have

$$\mathbf{P}(I) = \mathbf{P}(I \cap \oplus) + \mathbf{P}(I \cap \ominus) \tag{3.10}$$

We now express each of the probabilities $\mathbf{P}(I \cap \oplus)$, $\mathbf{P}(I \cap \ominus)$ in terms of the known ones: from Bayes's formula (3.9), we obtain

$$\mathbf{P}(I \cap \oplus) = \mathbf{P}(I|\oplus)\mathbf{P}(\oplus),$$

and

$$\mathbf{P}(I \cap \ominus) = \mathbf{P}(I|\ominus)\mathbf{P}(\ominus)$$

Substituting these into the right-hand side of (3.10), we obtain

$$\mathbf{P}(I) = \mathbf{P}(I|\oplus)\mathbf{P}(\oplus) + \mathbf{P}(I|\ominus)\mathbf{P}(\ominus) \tag{3.11}$$

3.4 Random variables

Let S be a sample space with a probability measure $\mathbf{P}$. A *random variable* is a function that assigns to each outcome of the experiment some numerical value. (In the terminology of section 1.7, a random variable is a function from S to the set of real numbers.)

Example 3.6 (a) *If two distinguishable dice are tossed simultaneously, the outcome is an ordered pair of scores, (s_1, s_2). The function that maps each of these outcomes to the sum of the two scores is a random variable.*

(b) *Suppose the experiment consists of recording all the vehicular traffic that has entered New York city over a 10-day period. Then the number of sedan cars among that traffic is a random variable.*

(c) *If the experiment consists of playing a game of Blackjack, a player's final score is a random variable.*

(d) *In the experiment of recording the clicks on an Internet ad during, say, six months, the total number of clicks is one random variable, and the total net profit from purchases on that ad is another.*

(e) *In the experiment consisting of purchasing a lawn mower of a given make and model and using the lawn mower until it first breaks, the time it took to break is a random variable.*

(f) *Let the experiment consist in selling merchandise in a store for one day. Then the list of all merchandise sold in a day is an outcome, and the set of all possible outcomes is the corresponding sample space. The total net profit for each possible outcome is a random variable.*

Exercise 3.9 *For each of the random variables described in example 3.6, describe the corresponding sample space, list at least three of the outcomes, and compute their respective values of the random variable.*

Exercise 3.4, above, illustrates that choosing a sample space suitable for the given problem can be far from obvious. The same is true of choosing helpful random variables, even for the apparently simple business problem in the following example.

Example 3.7 *A coffee shop accepts payment by credit card. The customer's usual way to approve the amount charged to the card is to sign the slip. Without this signature, the shop runs the risk that the customer will later try to reverse the charge under some false pretext. For the coffee shop, to deal with such a reversal–by verifying* *the validity of the charge–is to incur a* cost of verification, *which is assumed the same for all attempted reversals.*

On the other hand, to get the signature, the clerk must take additional time to print the slip and have the customer sign. This additional time also translates into a cost, called cost of delay, *which is assumed the same for all purchases.*

How can the shop manager use random variables to decide the minimal amount of a credit card purchase for which to require the customer's signature?

One possible approach is outlined in appendix B. Whatever approach is used, however, it involves computing the expected value *of a random variable, a concept introduced in section 3.5.1.*

3.5 Statistics of random variables

One often needs to characterize, using as few numbers as possible, some aspect in the behavior of a random variable. Such a number is often called a *statistic*. In this section, we present two such statistics, *expected value* (also known as *mean value* or *mathematical expectation*) and *variance*.

3.5.1 The expected (mean) value of a random variable

A widely encountered statistic is the one that reflects the "average behavior" of a random variable; for example, average profit from an Internet ad, or the average life expectancy of an appliance. To derive a clear concept of this statistic, called the *expected value* (also called the *mean value* or the *mathematical expectation*) of a random variable, we use specific examples and the concept of relative frequency introduced above.

The expected value of a random variable f on a sample space $S = \{T, H\}$

Imagine a game consisting in tossing a (possibly biased) coin, and getting a payoff accordingly as the toss yields *Heads* or *Tails*. The sample space is $S = \{T, H\}$, and suppose the probabilities $\mathbf{P}(T)$ and

$\mathbf{P}(H)$ are known. The mentioned payoff is then a random variable f; i.e., $f(T)$ and $f(H)$ are the respective payoffs for T and H.

After performing some N trials of the experiment (i.e., N tosses), we obtain N outcomes,

$$x_1, x_2, \ldots, x_N, \tag{3.12}$$

each being either T or H. Evaluating f for each of them, we obtain the values

$$f(x_1), f(x_2), \ldots, f(x_N) \tag{3.13}$$

The average of these values is

$$\frac{1}{N} [f(x_1) + f(x_2) + \ldots + f(x_N)] \tag{3.14}$$

Let N_T and N_H be the respective numbers of occurrences of T and H among the outcomes (3.12). The sum

$$f(x_1) + f(x_2) + \ldots + f(x_N)$$

is then equal to

$$f(T)N_T + f(H)N_H$$

Dividing the latter sum by N, we obtain that (3.14) equals

$$f(T)\frac{N_T}{N} + f(H)\frac{N_H}{N},$$

which has the form

$$f(T)(\text{relative frequency of } T) + f(H)(\text{relative frequency of } H) \tag{3.15}$$

In (3.15), each summand equals the value of f for an outcome multiplied by the relative frequency of that outcome. Since relative frequencies are approximations of probabilities, we replace the frequencies by the probabilities and obtain the following. *The expected value of a random variable f on a two-outcome sample space $S = \{T, H\}$ is defined as the quantity*

$$f(T)\mathbf{P}(T) + f(H)\mathbf{P}(H) \tag{3.16}$$

Exercise 3.10 *Use a fair coin (or a spreadsheet) to carry out ten trials of the experiment of tossing a fair coin, to obtain some ten outcomes*

$$x_1, x_2, \ldots, x_{10}$$

For each of the definitions of the payoff random variable f listed below, compute the values

$$f(x_1), f(x_2), \ldots, f(x_{10})$$

and calculate and compare the two quantities

$$\frac{1}{N} \left[f(x_1) + f(x_2) + \ldots f(x_{10}) \right] \tag{3.17}$$

and

$$f(T)\mathbf{P}(T) + f(H)\mathbf{P}(H) \tag{3.18}$$

(Since the coin is fair, we know that the two outcomes are equiprobable: $\mathbf{P}(T) = \mathbf{P}(H) = \frac{1}{2}$.)

(a) $f(T) = 1$, $f(H) = 1$

(b) $f(T) = 2$, $f(H) = 3$

(c) $f(T) = 120$, $f(H) = 4$

(d) $f(T) = -5$, $f(H) = 5$

(e) $f(T) = -3$, $f(H) = 9$

Repeat the exercise with more than ten trials (say, with twenty trials).

Exercise 3.11 *Do exercise 3.10 for a biased coin (simulated in a spreadsheet).*

Exercise 3.12 *Suppose a random variable f is defined for a three-outcome sample space $S = \{a, b, c\}$ with probability measure $\mathbf{P}$. If the experiment is repeated N times, to obtain N outcomes*

$$x_1, x_2, \ldots, x_N, \tag{3.19}$$

show that the average of the values

$$f(x_1), f(x_2), \ldots, f(x_N)$$

(given by formula (3.15)) equals

$$\left.\begin{array}{l} f(a)\,(\textit{relative frequency of } a) + \\[2mm] f(b)\,(\textit{relative frequency of } b) + \\[2mm] f(c)\,(\textit{relative frequency of } c) \end{array}\right\} \qquad (3.20)$$

Replacing the relative frequencies in (3.20) by the corresponding probabilities, we obtain that the expected value of the random variable f is

$$f(a)\mathbf{P}(a) + f(b)\mathbf{P}(b) + f(c)\mathbf{P}(c)$$

The expected value of a random variable with an arbitrary finite sample space

If f is a random variable on a finite sample space $S = \{x_1, \ldots, x_k\}$ with probability measure $\mathbf{P}$, the following definition is analogous to those obtained above for two- and three-outcome sample spaces. *The expected value of f, denoted by $E(f)$, is defined as*

$$E(f) = f(x_1)\mathbf{P}(x_1) + f(x_2)\mathbf{P}(x_2) + \ldots + f(x_k)\mathbf{P}(x_k) \qquad (3.21)$$

Exercise 3.13 *Find the expected value of the score when tossing*

(a) *a fair die.*

(b) *a die biased so that*

$$\mathbf{P}(1) = \frac{2}{17}, \ \mathbf{P}(2) = \frac{5}{17}, \ \mathbf{P}(3) = \frac{2}{17},$$

$$\mathbf{P}(4) = \frac{4}{17}, \ \mathbf{P}(5) = \frac{1}{17}, \ \mathbf{P}(6) = \frac{3}{17}$$

Exercise 3.14 *Find the expected value of the total score when toss-ing two fair dice.*

Exercise 3.15 *Suppose f and g are random variables (on the same sample space S and probability measure $\mathbf{P}$). One can then define the random variable $f + g$, which assigns to an outcome x the value*

$$f(x) + g(x)$$

If c is a real number, then cf is the random variable that assigns to an outcome x the value

$$c \cdot f(x)$$

The random variable $f \cdot g$ is defined analogously.

Use specific examples of S and $\mathbf{P}$ to test each of the following statements. For each statement, determine whether it holds true in general.

(a) *If a random variable is constant (i.e., has the same value for all outcomes), then this value is the expected value of the random variable.*

(b) *The expected value of the sum of two random variables is the sum of their individual expected values:*

$$E(f + g) = E(f) + E(g)$$

(c) *The expected value of a random variable multiplied by a constant equals to the variable's expected value multi-plied by that constant:*

$$E(c \cdot f) = c \cdot E(f)$$

(d) *The expected value of the product of two random vari-ables is the product of their individual expected values:*

$$E(fg) = E(f)E(g)$$

Hint: *Take $S = \{a, b\}$, $\mathbf{P}(a) = \mathbf{P}(b) = \frac{1}{2}$, and*

$$f(a) = -1, \ f(b) = 1, \quad g(a) = -1, \ g(b) = 1$$

Exercise 3.16 *A drunkard begins to walk on a coordinate line. Starting at the origin, he makes one step per second, stepping with the same probability (1/2) to the left or to the right. After N seconds, what is the expected position of the drunkard (i.e., how many steps away from, and on which side of) the origin?*

3.5.2   Variance

Consider the probability space $S = \{a, b, c\}$ with the probability measure **P** assigning probability $1/3$ to each outcome. The two random variables f and g, defined by

$$f(a) = 0, f(b) = 0, f(c) = 0; \quad g(a) = -1, g(b) = 1, g(c) = 0,$$

and depicted in the figure below (the stars represent the values of f; the shaded circles, the values of g), have the same expected value.

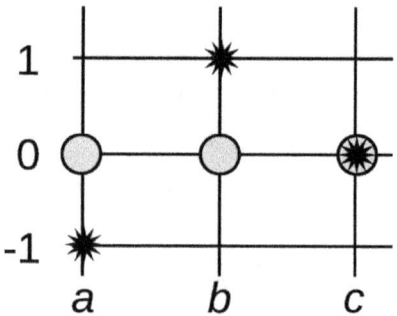

Yet, the behaviors of f and g are qualitatively different: one is constantly 0, the other varies from -1 to 1. Thus, the statistic *expected value*, introduced in the previous section, fails to capture this difference in behavior.

A statistic that will succeed at this task should capture, for each random variable, the distances between this variable's expected value and its values for individual outcomes. For f, these distances are

$$|f(a) - E(f)|, \ |f(b) - E(f)|, \ |f(c) - E(f)|$$

For technical reasons beyond the scope of this book, however, it is often more convenient to use, instead, the squared distances

$$(f(a) - E(f))^2, \ (f(b) - E(f))^2, \ (f(c) - E(f))^2$$

Since such a square is assigned to each outcome, we have in fact constructed a new random variable, which we denote by F:

$$F(a) = (f(a)-E(f))^2, \ F(b) = (f(b)-E(f))^2, \ F(c) = (f(c)-E(f))^2,$$

The random variable F measures the squared difference between f and the expected value of f, i.e. measures how dispersed the values of f are about its expected value. The expected value of F is called, *the variance of f* (also called *the dispersion of f*). Similarly for g.

Exercise 3.17 *Calculate the variances of the random variables f and g defined at the beginning of this section.*

In general, if f is a random variable on a finite sample space $S = \{x_1, \ldots, x_k\}$ with probability measure $\mathbf{P}$, the *variance of f*, denoted $V(f)$, is defined as

$$
\begin{aligned}
V(f) \ &= \ [f(x_1) - E(f)]^2\mathbf{P}(x_1) \\
&+ \ [f(x_2) - E(f)]^2\mathbf{P}(x_2) \\
&+ \ \ldots \\
&+ \ [f(x_k) - E(f)]^2\mathbf{P}(x_k)
\end{aligned}
$$

In sketchy but concise prose, "the variance is the expected squared deviation from the mean."

Exercise 3.18 *What is the variance of the fair die score?*

3.6 If you need further knowledge of probability

The material and exercises in this chapter are minimal. Problem book [12] is an excellent way to test your understanding of the ideas set forth here.

A more advanced study of probability theory leads to sample spaces where some events are not *observable*.

Example 3.8 *Consider five beakers, three containing different acids, and the other two, different bases. The experiment consists in picking a beaker at random and putting into it a drop of litmus[5]. The sample space can be written*

$$S = \{a_1, a_2, a_3, b_1, b_2\}$$

The described experiment can help us distinguish the event $A = \{a_1, a_2, a_3\}$ from the event $B = \{b_1, b_2\}$, but it cannot help distinguish, for example, the event $\{a_1, a_2\}$ from the event $\{a_3\}$. The events $A, B, \emptyset$ and S are said to be observable *($\emptyset$ and S are always observable); all other events,* unobservable.

An exposition of probability theory which includes observability of events can be found in [15]. Another complication that will be added by this, more advanced, study is the occurrence of sample spaces that have infinitely many outcomes (these did arise in the present book on occasion, e.g. in the last part of exercise 3.4, but were not the focus).

The presence of unobservable events entails further restrictions that need to be placed on random variables: the event that a random variable equals a given value may be unobservable. The apparatus needed here is that of *measure theory* (see the latter part of [17]), which requires, first, a solid course in mathematical analysis.

One of the many engineering disciplines adjacent to probability theory is *information theory*. A pleasant first acquaintance with it can be made using [14]; a first elementary study, using [6].

---

[5]Litmus is a chemical that turns a different color accordingly as the fluid in a vessel is a base or an acid.

Bibliography

[1] V. I. Arnold, *On teaching mathematics* (article) 1997.

[2] V. B. Alekseev, *Abel's Theorem in Problems and Solutions: Based on the lectures of Professor V.I. Arnold* (book).

[3] I. Gelfand, A. Shen, *Algebra*, Birkhäuser Boston, 1993.

[4] I. Gelfand et. al., *Method of Coordinates*.

[5] K. Gödel. *On Formally Undecidable Propositions*.

[6] A. Khinchin. *Mathematical Foundations of Information Theory*, Dover, 1957.

[7] A. P. Kiselev. *Kiselev's Geometry. Book I. Planimetry*, Sumizdat, 2006.

[8] A. N. Kolmogorov. *Foundations of the theory of probability*.

[9] D. Knuth, *The Art of Computer Programming, Volume 1*, Addison-Wesley, 1997

[10] Z. Manna. *Mathematical theory of computation*.

[11] B. Martin. *Statistics for physicists*.

[12] F. Mosteller, *Fifty Challenging Problems in Probability with Solutions*, Dover, 1987.

[13] O. Ore, *Graphs and their uses*, Cambridge University Press, 1990.

[14] A. Rényi, *A Diary on Information Theory.*

[15] A. Rényi, *Probability Theory*, Dover, 2007.

[16] B. L. Van der Waerden. *Mathematical Statistics.*

[17] B. Vulikh, *A Brief Course in the Theory of Functions of a Real Variable.*

Appendix A

Gauss's summation formula

The sum of the first N positive integers, i.e. the value

$$s = 1 + 2 + \ldots + N,$$

can be computed using the following approach. Rewriting the above sum in reverse order,

$$s = N + (N - 1) + \ldots + 1,$$

and then the two sums underneath each other in neat columns,

$$s = 1 + 2 + \ldots + N$$
$$s = N + (N - 1) + \ldots + 1$$

we add columnwise:

$$2s = (N + 1) + (N + 1) + \ldots + (N + 1)$$

On the right-hand side, the expression $(N + 1)$ occurs N times, so

$$2s = (N + 1)N$$

Dividing by 2, we obtain *Gauss's*[1] *summation formula*

$$1 + 2 + \ldots + N = \frac{1}{2}(N + 1)N \qquad (A.1)$$

---

[1]Carl Friedrich Gauss (1777-1855) was a German mathematician and physicist.

Appendix B

The coffee shop problem

One approach to the problem stated in example 3.7 is as follows. Let v and d denote, respectively, the cost of verification and the cost of delay.

Each credit card transaction without a signature is regarded as an experiment. Each outcome is recorded by writing down the following information: the immediate profit from the purchase, and whether the customer later (say, within a week) tries to reverse the charge. Accordingly, an outcome with profit $\$x$ is written $R\$x$ if a reversal is later attempted, and $A\$x$ otherwise (with A for *approved*). The sample space S consists of all $R\$x, A\$x$. For example, a \$4.57 charge with no reversal attempted is outcome $A\$4.57$.

The random variables needed here are two: the final profit from a signed transaction, and the final profit from an unsigned transaction. They will be denoted, respectively, by f_s and f_u. Each has the form

$$(\text{immediate profit}) - (\text{cost of delay}) - (\text{cost of verification})$$

For a signed transaction, the cost of delay is d, and the cost of verification is zero. Thus,

$$\left. \begin{array}{rcl} f_s(A\$x) & = & x - d \\ f_s(R\$x) & = & x - d \end{array} \right\}$$

For an unsigned transaction, the cost of delay is zero, and the cost of verification is zero or v, accordingly as a reversal later is or is not attempted. Thus,

$$\left.\begin{array}{rcl} f_u(A\$x) &=& x \\ f_u(R\$x) &=& x - v \end{array}\right\}$$

Thus, one wants to choose the minimum purchase amount to require a signature so that, for that amount, the profits f_u and f_s are "on the average" the same. The behavior of a random variable "on the average" is captured by the concept of *expected value*, introduced in section 3.5.1.

Appendix C

Newton's binomial formula

An expression equal to a sum of two terms is called *a binomial*[1].
The following theorem tells how to compute quickly, using Pascal's
Triangle (exercise 2.12 in section 2.7), a power of a binomial.

Theorem C.1 *If x and y are real numbers, and if n is a nonnegative integer, then*

$$(x+y)^n = y^n + C_1^n x^1 y^{n-1} + C_2^n x^2 y^{n-2} + \ldots + C_{n-1}^n x^{n-1} y^1 + x^n \quad (C.1)$$

Note that, in the above sum, the exponent of x matches the subscript
of C, and that the exponents of x and y in each term add up to n.
Formula (C.1) is called, *Newton's binomial formula*.

    Proof of theorem C.1. To prove this formula, one writes the
power $(x+y)^n$ as the product

$$(x + y) \cdot (x + y) \cdot \ldots \cdot (x + y), \quad (C.2)$$

the factor $(x + y)$ occurring n times, and reasons as follows for each
$k = 0, 1, \ldots, n$.

    In the right-hand side of (C.1), the k-th summand,

$$C_k^n \, x^k \, y^{n-k},$$

---

[1]The Latin root *bi* means two, as in *bifocal glasses* and *bicentennial release*.

is obtained by multiplying out the product (C.2) and collecting the like powers

$$x^k y^{n-k}$$

Each such power is, in turn, obtained by choosing, from each of the n factors in (C.2), either the term x or the term y, so that x is chosen from k of the factors, and y from the remaining ones. The obtained sequence of x's and y's has length n, and contains k occurrences of x and $(n-k)$ occurrences of y. Multiplying the terms of the sequence, we get

$$x^k y^{n-k}$$

Since (by theorem 2.11) there are C_k^n ways to choose such a sequence, collecting the like powers will produce C_k^n terms of the form $x^k y^{n-k}$. Collecting them, one obtains the summand $C_k^n x^k y^{n-k}$ in (C.1). This completes the proof.

Because of theorem C.1, the numbers C_n^k are called *binomial coefficients*.

Appendix D

Hints, answers, and solutions to selected exercises

- **1.1** Yes.

- **1.2**

    (a) Yes.

    (b) Yes.

    (c) Yes.

    (d) No.

    (e) Yes.

    (f) No.

- **1.4**

    (a) Record the membership functions m_R, m_S, and $m_{R \cap S}$ in rows 1, 2, 3, respectively:

	A	B	C	D	E	F		
1	0	1	0	1	1	1	⋮	⋮
2	0	1	1	1	1	0	⋮	⋮
3	= A1 * A2	= B1 * B2	= C1 * C2	= D1 * D2	= E1 * E2	= F1 * F2	⋮	⋮
⋅								
⋅							⋮	⋮

Here the notation B3, for instance, means "the contents of the cell in column B, row 3."

(b) $m_{R \cup S}(x) = $ the larger of $m_R(x), m_S(x)$

(c) $m_{Q \setminus R}(x) = 1 - m_R(x)$

- **1.5**

  (a) The truth is

$r(x)$	$s(x)$	$(r \wedge s)(x)$
F	F	F
F	T	F
T	F	F
T	T	T

  The analogy with set operations is that $(r \wedge s)(x)$ is true for those x, and only for those, that are elements of $R \cap S$.

- **1.7**

  (a) $X \times Y = \{(\heartsuit, a), (\heartsuit, c), (\heartsuit, j), (\spadesuit, a), (\spadesuit, c), (\spadesuit, j), (\diamondsuit, a), (\diamondsuit, c), (\diamondsuit, j)\}$

  (b) $X \times Y = \{(\heartsuit, \alpha), (\heartsuit, a), (\heartsuit, c), (1, \alpha), (1, a), (1, c), (a, \alpha), (a, a), (a, c)\}$

  (c) $X \times Y = \{(m, n) : m \text{ and } n \text{ are integers}\}$

  (d) $X \times Y = \emptyset$

  (e) $X \times Y = \emptyset$

- **1.8**

  (a) (*Disclaimer:* The following choice of five 4-tuples is purely fictional and does not reflect any real sandwiches or the author's culinary skill.)

  $$(\mathtt{wheat}, \mathtt{ketchup}, \mathtt{onion}, \mathtt{swiss}),$$
  $$(\mathtt{wheat}, \mathtt{ketchup}, \mathtt{onion}, \mathtt{cheddar}),$$
  $$(\mathtt{rye}, \mathtt{mayonnaise}, \mathtt{avocado}, \mathtt{smoked}),$$
  $$(\mathtt{wheat}, \mathtt{mayonnaise}, \mathtt{cucumber}, \mathtt{swiss}),$$
  $$(\mathtt{rye}, \mathtt{mayonnaise}, \mathtt{tomato}, \mathtt{feta})$$

- **1.9** No. Consider the following shopping list:

Item	Quantity
milk	2 gallons
apple juice	3 bottles
eggs	1 dozen
green beans	1.5 lbs

The last quantity is not an integer.

- **1.10**

   (a) $f(1) = 2, f(3) = 6, f(4) = 8$.

   (b) Yes: this f satisfies the definition of a function (the first sentence of section 1.7).

   (c) Yes: this f satisfies the definition of a function (the first sentence of section 1.7).

   (d) No: this f does not map 3 to any element in Y.

- **1.11**

   (a) No: this function does not map a to any element of Y, hence is not a function from X to Y.

   (b) No: this function maps b and c to the same element of Y. (Also, this function fails to map anything to the element 1 of Y.)

   (c) No: this is not a function from X to Y, as it attempts to map a to two different elements of Y. A 1-to-1 correspondence, by definition (section 1.8), must be a function.

   (d) No: this function does not map c to any element of Y, hence is not a function from X to Y.

   (e) No: this is not a function from X to Y.

   (f) Yes.

- **1.12**

(a) No. If there was such a 1-to-1 correspondence (denote it by f), then

$$f(a), f(b), f(c), f(d)$$

would be four distinct elements of $\{1, 2, 3\}$, which is impossible since the latter set has only three elements.

(b) No. If there was such a 1-to-1 correspondence (denote it by f), then by listing the three elements

$$f(a), f(b), f(c)$$

of the set $\{1, 2, 3, 4\}$ we would have listed all elements of the latter set, which is impossible because the set has four elements.

(c) No.

- **2.1** One experiment (of the possible many) that can be done is to take

$$X_1 = \{1, 2, 3, 4\}, \quad X_2 = \{a, b\}, \quad X_3 = \{\heartsuit, \diamondsuit, \spadesuit\}$$

The Cartesian product $X_1 \times X_2 \times X_3$ of these three sets can be visualized as follows:

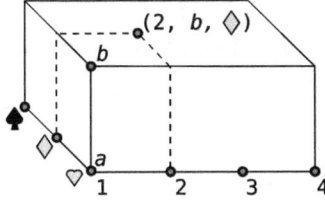 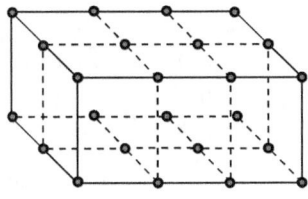

These figures suggest that by thinking of $|X_1|, |X_2|, |X_3|$ as, respectively, "the length," "the width," and "the height," one sees that the total number of elements in $X_1 \times X_2 \times X_3$ is "the volume." Similar reasoning (but without the 3-dimensional visualization) applies to Cartesian products of more than three sets.

- **2.2**

(a) Since the elements of $X \cap Y$ are found both in X and in Y, these elements get counted twice in the sum $|X| + |Y|$. For example, if $X = \{a, b, c, d, e\}, Y = \{c, d, e, f, g, h\}$, then

$$|X| + |Y| = 5 + 6 = 11$$

In the latter sum, each of the five elements a, b, f, g, h is counted once, while each of the elements c, d, e of $X \cap Y$ is counted twice:

$$11 = 5 + 2(3)$$

To correct for this, one must subtract "the second copy" of $|X \cap Y|$, obtaining:

$$|X \cup Y| = |X| + |Y| - |X \cap Y| \qquad \text{(D.1)}$$

Another formula for $|X \cup Y|$ is suggested by the figures

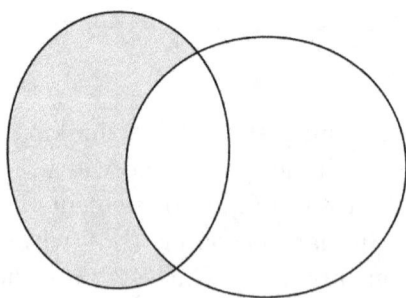

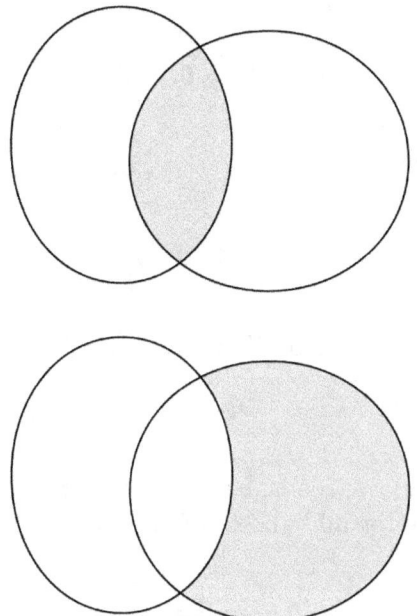

This formula is

$$|X \cup Y| = |X \setminus Y| + |X \cap Y| + |Y \setminus X| \qquad \text{(D.2)}$$

(b) Let the set X_2 consist of all the multiples of 2 between 0 and 480, and construct X_3 similarly. The set of all numbers divisible either by 2, or by 3, or by both, is $X_2 \cup X_3$. By formula (D.1),

$$
\begin{aligned}
|X_2 \cup X_3| &= |X_2| + |X_3| - |X_2 \cap X_3| \\[2mm]
&= \tfrac{480}{2} + \tfrac{480}{3} - \tfrac{480}{6} \\[2mm]
&= 240 + 160 - 80 \\[2mm]
&= 320
\end{aligned}
$$

- **2.3** By the definition of factorial (section 2.3), one has

$$0! = 1, \quad 1! = 1$$

Using recurrence (2.4), one obtains:

$$2! = 1!2 = 2, \quad 3! = 2!3 = 6, \quad 4! = 3!4 = 24, \quad 5! = 4!5 = 120$$

- **2.4** *Answer:*
$$\frac{127!}{8!}$$

- **2.6**

  (a) The subsets of $X = \{a, b\}$ are:

  $$\emptyset, \{a\}, \{b\}, \{a, b\}$$

  (b) Using a spreadsheet, we record the membership functions for all subsets of $X = \{a, b, c\}$:

	A	B	C	
1	0	0	0	...
2	0	0	1	...
3	0	1	0	...
4	0	1	1	...
5	1	0	0	...
6	1	0	1	...
7	1	1	0	...
8	1	1	1	...

  Consequently, X has 8 subsets.

  (c) If Y is a subset of X, then the membership function m_Y is a function from X to $\{0, 1\}$. Conversely, every function from X to $\{0, 1\}$ is the membership function of some subset of X. It follows that there are as many subsets of X as there are membership functions. By theorem 2.3, there are $2^{|X|}$ of them.

- **2.7** *Answer:*
$$\frac{26!}{22!}$$

- **2.8** Every such password is a sequence of 5 distinct terms from the set

$$A = \{a, b, \ldots, z, !, \#, @, \_\} \quad (n = |A| = 30)$$

Conversely, every such sequence is a permissible password. Thus, there are as many passwords as there are sequences. By theorem 2.4, the total number of such sequences is

$$\frac{n!}{(n-k)!} = \frac{30!}{25!}$$

- **2.9** *Answer:*

$$\frac{(16+27)!}{16! \, 27!}$$

- **2.10**

  (a) *Answer:* C_6^{100}

  (b) Once a Sneaker is chosen (one way to accomplish this), five more bars must be chosen from among the remaining 99. *Answer:* C_5^{99}.

- **2.11**

  (a) Every such expression is obtained by first writing

  $$1+1+1+1+1+1+1+1+1+1, \qquad \text{(D.3)}$$

  and then arbitrarily choosing and circling three of the +'s. For instance, writing

  $$1+1 \oplus 1+1+1 \oplus 1+1+1+1 \oplus 1$$

  corresponds to taking $a_1 = 2, \quad a_2 = 3, \quad a_3 = 4, \quad a_4 = 1$, i.e.

  $$\underbrace{1+1}_{a_1} \quad \oplus \quad \underbrace{1+1+1}_{a_2} \quad \oplus \quad \underbrace{1+1+1+1}_{a_3} \quad \oplus \quad \underbrace{1}_{a_4}$$

  Thus, the question is, *How many ways are there to choose three of the nine pluses in (D.3)? Answer:* C_3^9.

(b) If $a_k \geq 0$, then $a_k + 1 > 0$. Now take $b_k = a_k + 1$ and note that we must have

$$
\begin{aligned}
b_1 + b_2 + b_3 + b_4 &= (a_1 + 1) + (a_2 + 1) + (a_3 + 1) + (a_4 + 1) \\
&= a_1 + a_2 + a_3 + a_4 + 4 \\
&= 14
\end{aligned}
$$

Answer: C_3^{13}

(c) *Answer:* C_{k-1}^{n-1}

(d) 1. Put a marble in each jar; this guarantees that no jar is empty and leaves $(n-k)$ marbles. Now count the ways of distributing the remaining marbles. The number of these distributions is the same as the number of ways to write $(n-k)$ as a sum of k nonnegative integers:

$$n - k = a_1 + a_2 + \ldots + a_k$$

Taking $b_1 = a_1 + 1, b_2 = a_2 + 1, \ldots, b_k = a_k + 1$, we obtain

$$b_1 + b_2 + \ldots + b_k = a_1 + a_2 + \ldots + a_k + k = n$$

Answer: C_{k-1}^{n-1}

2. *Hint:* First choose a set of m jars to be kept empty, then put a marble in each of the remaining $(k-m)$ jars, and distribute the remaining marbles among those $(k-m)$ jars.

- **2.12**

(a)

```
            1
          1   1
        1   2   1
      1   3   3   1
    1   4   6   4   1
  1   5  10  10   5   1
```

(b) The pyramid is symmetric with respect to the vertical line through the tip of the pyramid:

$$C_k^n = \frac{n!}{(n-k)!k!} = \frac{n!}{(n-k)!(n-(n-k))!} = C_{n-k}^n$$

(c) *Partial answer:* The first five rows of the Triangle are computed as follows:

	A	B	C	D	E	F	
1	1	0	0	0	0	0	...
2	1	1	0	0	0	0	...
3	1	= A2 + B2	1	0	0	0	...
4	1	= A3 + B3	= B3 + C3	1	0	0	...
5	1	= A4 + B4	= B4 + C4	= C4 + D4	1	0	...

- **3.1**

  (a) $E \cap F$

  (b) $E \cup F$

  (c) $E \setminus F$

  (d) $S \setminus E$

- **3.3**

  (a) *Indication:* To simulate this bias for a given N, copy the formula

  $$= 1 * (\mathbf{rand}(1) > 0.2)$$

  into the leftmost N cells in the top row.

  (b) *Indication:* To simulate this bias for a given N, copy the formula

  $$= 1 * (\mathbf{rand}(1) > 2/3)$$

  into the leftmost N cells in the top row.

- **3.4**

(a) Sample space: $S = \{TT, TH, HT, HH\}$. Two possible different probability measures:

x	TT	TH	HT	HH
$\mathbf{P}(x)$	1/4	1/4	1/4	1/4

for a fair coin, and

x	TT	TH	HT	HH
$\mathbf{P}(x)$	1/9	2/9	2/9	4/9

for a biased coin. (There are others.)

(b) *Partial answer:* The sample space consists of the six possible scores that may result from a toss: $S = \{1, 2, 3, 4, 5, 6\}$. If the die is fair, the corresponding probability measure makes all six outcomes equally probable, hence assigns to each outcome the probability of 1/6. A choice of a probability measure for a biased die is left to the reader.

(c) *Partial answer:* The sample space is

$$S = \{TTT, TTH, THT, THH, HTT, HTH, HHT, HHH\}$$

The task of choosing two different probability measures is left to the reader.

- **3.5**

    (a) True. Since events E and E^c are mutually exclusive, one has

    $$\mathbf{P}(E^c) + \mathbf{P}(E) \quad = \mathbf{P}(E^c \cup E) \qquad \text{(by (3.6c))}$$

    $$= \mathbf{P}(S)$$

    $$= 1 \qquad \text{(by (3.6a))}$$

    i.e.

    $$\mathbf{P}(E^c) + \mathbf{P}(E) = 1$$

(b) True. Since events $(F \cap E)$ and $(F \cap E)^c$ are mutually exclusive, and since their union equals F, equality (3.6c) applies. A pictorial illustration is

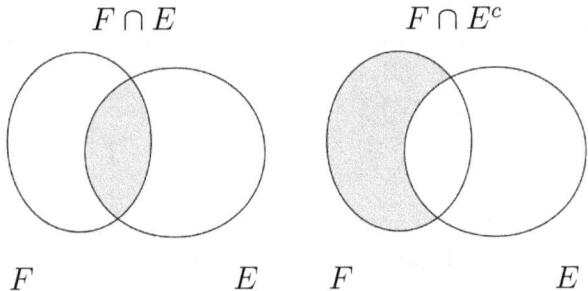

$F \cap E$ $\qquad\qquad$ $F \cap E^c$

F $\qquad\qquad$ E $\qquad$ F $\qquad\qquad\qquad$ E

(c) True. If $E \subset F$, let $A = F \setminus E$.

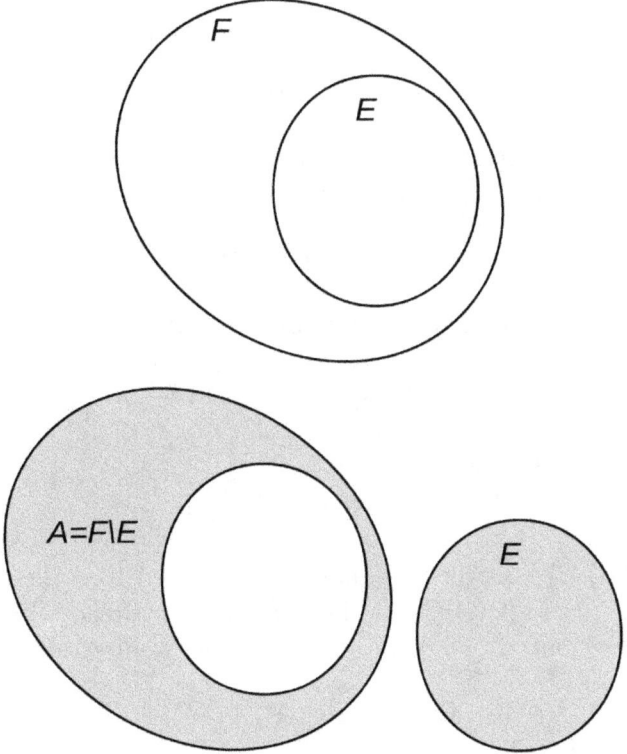

Then the events A and E are mutually exclusive, and

$A \cup E = F$, so (by (3.6) again) one has

$$\begin{aligned} \mathbf{P}(F) = \mathbf{P}(A \cup E) &= \mathbf{P}(A) + \mathbf{P}(E) \\ &\geq \qquad\quad \mathbf{P}(E) \end{aligned}$$

(d) False. A counterexample is a case where $\mathbf{P}(A) = 0$. (Example 3.3 shows that this can happen even if A is nonempty.)

- **3.8**

(a) Since there are four queens in the deck and since the cards are drawn randomly (i.e., all have equal probability of being drawn), event A has probability

$$\mathbf{P}(A) = \frac{4}{52} = \frac{1}{13}$$

Since exactly one fourth of the deck are spades,

$$\mathbf{P}(B) = \frac{1}{4}$$

Event $A \cap B$ is the event that a queen of spaces is drawn, which equals

$$\mathbf{P}(A \cap B) = \frac{1}{52}$$

Consequently, by definition (3.8),

$$\mathbf{P}(A|B) = \frac{1/52}{1/4} = \frac{4}{52} = \frac{1}{13},$$

and

$$\mathbf{P}(B|A) = \frac{1/52}{1/13} = \frac{13}{52} = \frac{1}{4},$$

(b) To answer this question, we need to compare $\mathbf{P}(A)$ and $\mathbf{P}(A|C)$. The former has already been computed. To compute the latter, we first obtain

$$\mathbf{P}(C) = \frac{1}{2}, \quad \mathbf{P}(A \cap C) = \frac{2}{52} = \frac{1}{26}$$

The computation of $\mathbf{P}(A|C)$ using (3.8) and comparison between $\mathbf{P}(A)$ and $\mathbf{P}(A|C)$ are left to the reader.

- **3.9**

  (a) The sample space is the set of all ordered pairs of die scores:

  $$S = \{1, 2, \ldots, 6\} \times \{1, 2, \ldots, 6\}$$

  $$= \{(1, 1), (1, 2), \ldots, (5, 6), (6, 6)\}$$

  Three of the outcomes are:

  $$x_1 = (1, 1), \quad x_2 = (3, 4), \quad x_3 = (5, 2)$$

  Denoting the random variable by f, we have

  $$f(x_1) = 1+1 = 2, \quad f(x_2) = 3+4 = 7, \quad f(x_3) = 5+2 = 7$$

  (b) If there exist a total of N vehicle types (one of them being sedan), each outcome is a table of the form

	type 1	type 2	...	type N
# vehicles entering NYC				

  the bottom row filled with nonnegative integers. In other words, such an outcome can be described as a function from the set T of all vehicle types to the set $\{0, 1, \ldots\}$ of all nonnegative integers.

  The sample space S is the set of all such tables (or, equivalently, of all such functions).

  Three of the outcomes are:

  - an outcome when 4 sedans entered the City
  - an outcome when 7 sedans entered the City
  - an outcome when 0 sedans entered the City

  The respective values of the random variable in question are $4, 7, 0$.

- **3.10** Let us suppose that the ten trials carried out have yielded the outcomes

  $$THTTHTHHTT$$

(a) The quantities (3.17) and (3.18) turn out to be, respectively,

$$\frac{1}{10}(1+\ldots+1) = 1$$

and

$$\frac{1}{2}1 + \frac{1}{2}1 = 1$$

(b) The quantities (3.17) and (3.18) turn out to be, respectively,

$$\frac{1}{10}(2+3+2+2+3+2+3+3+2+2) = \frac{12}{5} = 2.4$$

and

$$\frac{1}{2}2 + \frac{1}{2}3 = \frac{5}{2} = 2.5$$

- **3.12** Let N_a, N_b, N_c be the respective numbers of occurrences of a, b, c among the outcomes (3.19). Consequently, in the average (3.15), N_a of the summands equal $f(a)$, N_b of the summands equal $f(b)$, and N_c of the summands equal $f(c)$. Therefore, (3.15) can be written

$$\frac{1}{N}\left(N_a f(a) + N_b f(b) + N_c f(c)\right) = f(a)\frac{N_a}{N} + f(b)\frac{N_b}{N} + f(c)\frac{N_c}{N}$$

The ratios N_a/N, N_b/N, N_c/N are the respective relative frequencies of a, b, c.

- **3.13**

  (a) Since with a fair die every score has the same probability, $1/6$, the expected value (defined by equality (3.21)) is

  $$1 \cdot \frac{1}{6} + 2 \cdot \frac{1}{6} + \ldots + 6 \cdot \frac{1}{6} = \frac{21}{6} = \frac{7}{2}$$

  (b) Definition (3.21) gives the expected value

  $$1 \cdot \frac{2}{17} + 2 \cdot \frac{5}{17} + 3 \cdot \frac{2}{17} + 4 \cdot \frac{4}{17} + 5 \cdot \frac{1}{17} + 6 \cdot \frac{3}{17};$$

  the computation is left to the reader (or to the reader's spreadsheet).

- **3.14** *Indication:* Fill a six-by-six table with the rows labeled $1, 2, \ldots, 6$, the columns also labeled $1, 2, \ldots, 6$, and the entry in row i and column j being equal to $(i+j)$. This entry is the value of the random variable corresponding to the outcome (i, j). Since each outcome is equally probable (hence has probability $1/36$), the expected value of the random variable in question is

$$\frac{1}{36} \cdot (\text{the sum of all the entries in the table})$$

The latter sum can be calculated, without computing technology, using Gauss's summation formula (A.1) in appendix C.

- **3.15**

  (a) True. If the random variable f takes the same value, say a, for all outcomes, then (3.21) evaluates to

  $$E(f) = a \cdot (\mathbf{P}(x_1) + \mathbf{P}(x_2) + \ldots + \mathbf{P}(x_k)) = a \cdot 1 = a$$

  (b) True. To see this, we first consider the special case when $S = \{x_1, x_2\}$. In this case, one has

  $$E(f + g)$$

  $$= \mathbf{P}(x_1)[f(x_1) + g(x_1)] + \mathbf{P}(x_2)[f(x_2) + g(x_2)]$$

  $$= \mathbf{P}(x_1)f(x_1) + \mathbf{P}(x_2)f(x_2) + \mathbf{P}(x_1)g(x_1) + \mathbf{P}(x_2)g(x_2)$$

  $$= E(f) + E(g)$$

  A similar algebraic manipulation with a general finite sample space $S = \{x_1, \ldots, x_N\}$ yields that the formula
  $$E(f + g) = E(f) + E(g)$$
  is generally true for finite sample spaces.

  (c) *Indication:* Use an approach analogous to that used in the previous part of the exercise. *Answer:* True.

- **3.17** *Answer:*
$$V(f) = 0, \quad V(g) = \frac{2}{3}$$

- **3.18** If f is the die score, then
$$E(f) = \frac{1}{6}(1 + \ldots + 6) = \frac{21}{6} = \frac{7}{2},$$

and so

$$V(f) = \frac{1}{6}\left[\frac{7}{2} - 1\right]^2 + \frac{1}{6}\left[\frac{7}{2} - 2\right]^2 + \ldots \frac{1}{6}\left[\frac{7}{2} - 6\right]^2;$$

the calculation is left to the reader.

Index